T0252020

MODEL INDUCTION FROM DATA: TOWARDS THE
COMPUTATIONAL ENGINES IN HYDRAULICS AND HYDROLOGY

For my father

Model Induction from Data: Towards the Next Generation of Computational Engines in Hydraulics and Hydrology

DISSERTATION

Submitted in fulfilment of the requirements of
the Board for Doctorates of Delft University of Technology
and of the Academic Board of the International Institute for Infrastructural, Hydraulic
and Environmental Engineering for the Degree of DOCTOR
to be defended in public
on Thursday, 28 February 2002 at 10:30 hours in Delft, The Netherlands

by

YONAS BERHAN DIBIKE

born in Addis Ababa, Ethiopia
Master of Science with Distinction, IHE

Taylor & Francis
Taylor & Francis Group
LONDON AND NEW YORK

This dissertation has been approved by the promoter:
Prof. dr. M.B. Abbott TU Delft/IHE Delft, The Netherlands

Members of the Awarding Committee:

Chairman	Rector Magnificus TU Delft, The Netherlands
Co-chairman	Rector IHE Delft, The Netherlands
Prof. dr. ir. P. van der Veer	TU Delft, The Netherlands
Prof. dr. ir. A.E. Mynett	TU Delft, The Netherlands
Prof. dr. R. Falconer	Cardiff University, United Kingdom
Prof. dr.-ing. K.P. Holz	BTU Cottbus, Germany
Dr. V. Babovic	Danish Hydraulic Institute, Denmark

Published by Taylor & Francis
2 Park Square, Milton Park, Abingdon, Oxon, OX14 4RN
270 Madison Ave, New York NY 10016

Transferred to Digital Printing 2007

ISBN 90 5809 356 5

Publisher's Note
The publisher has gone to great lengths to ensure the quality of this reprint but points out that some imperfections in the originalt may be apparent

Contents

Abstract

Two basic approaches could be distinguished to modelling the physical environment, namely *deductive* and *inductive* methods. Deduction is the application of general laws of nature to represent a particular physical phenomenon in mathematical or other systematic forms of representation, while induction is the synthesis of a general law or other form of representation from many particular observations of the phenomena in a given physical system. Even though the deductive approach is widely and successfully used in many fields of studies, its range of applicability is sometimes affected by such factors as incomplete understanding of the processes involved and loss of accuracy in discretising the governing equations for their numerical solution. Moreover, the computational time and resources required to simulate a real physical system based on the deductive approach is usually very high, making it difficult for such models to be incorporated as part of decision-support systems.

The main subject that is addressed in this thesis is, therefore, model induction from data for the simulation of hydrodynamic processes in the aquatic environment. First, some currently popular artificial neural network architectures are introduced, mainly as technical devices by means of which learning from data is made possible by describing input-output processes in terms of activation patterns defined over nodes in highly interconnected networks. It is then argued that these devices can be regarded as domain knowledge incapsulators by applying the method to the generation of wave equations from hydraulic data and showing how the equations of numerical-hydraulic models can, in their turn, be recaptured using artificial neural networks. The thesis also demonstrates how artificial neural networks can be used to generate numerical operators on non-structured grids for the simulation of hydrodynamic processes in two-dimensional flow systems by applying the technique to the simulation of tidal flows in an estuary. Moreover a methodology has been derived for developing generic hydrodynamic models using artificial neural network with schemes that can be applied over arbitrary bathymetries with variable distance and time steps. The thesis also highlights one other model induction technique, namely that of the support vector machine, as an emerging new method which could be considered as a generalisation of artificial neural networks with a potential to provide more robust models than those obtained from most other techniques. This has been demonstrated on two case studies by applying the technique to hydrologic and hydrodynamic modelling problems.

Acknowledgements

I would like to take this opportunity to thank the International Institute for Infrastructural Hydraulic and Environmental Engineering, IHE-Delft for financing the work in this thesis.

First of all, I would like to thank my supervisor and promoter Prof. Michael Abbott for creating the opportunity for me to do my Ph.D. research and his guidance throughout the work and during the writing of the thesis.

I would also like to express my thanks to all the people in the hydroinformatics core who contributed to this work in one way or another and made my stay at IHE quite pleasant and fruitful. A special word of thanks to Prof. Roland Price, Prof. Arthur Mynett, Dr Anthony Minns, Dr Dimitri Solomatine, Dr Vladan Babovic and Dr. Henk van der Boogaard for the valuable discussion and the support I got at various stages of this research work. I also wish to thank Ir Jan Luijendijk and Ms. Jos Bult for their help in preparing the Dutch version of the summary of this thesis.

Finally I would like to thank my family and friends for their continued moral support over the years.

Chapter 1

Introduction

Any physical system in the *real* world can be represented by a model to different degrees of accuracy with different forms of representation. The first types of such models in hydraulics are those simple empirical relationships that have served to study the performance of various hydraulic systems over a range of different circumstances. Similarly, a model could also be a verbal description of the cause-effect relations that are found in a system, such as in the form of rules (like | if ... then... | statements). Also, a model can itself be a smaller physical system that parallels the action of a large system as in the case of reduced scale models. At the beginning, scale models in most hydraulics laboratories were used as illustrative examples and as means of conducting simple experiments to estimate the values of some empirical coefficients. When the complexity and scope of large hydraulic structures began to present problems which could not be solved using empirical models, hydraulic scale models evolved into a more important role of providing quantitative and physically reliable results on which design decisions could be based (Cunge et al., 1980).

With the advent of digital computing technologies, mathematical modelling evolved strongly using equations to describe the quantitative relationship between different system parameters - even the behaviour of the whole system - based on the fundamental principles such as conservation of mass, momentum and energy. Nowadays, large proportions of simulation experiments are performed using mathematical models and the shift in modelling paradigms from scale models to mathematical models was already indicated by Cunge et al. (1980):

> As engineering projects become larger and economic considerations were more and more integrated into overall planning, scale models reached a natural limit to the scope of their application.... Even though methods like scale distortion and engineering experience were used to extend the scope of scale models, there came a point at which new techniques should be used to obtain simulations of natural processes which are reliable and economical. These *new* techniques are those of mathematical models

The great economic and commercial incentives to develop practical numerical modelling methods in hydraulics were also emphasised by Abbott (1979 // 1985):

> … This development naturally brings with it a very through reformulation of hydraulics to suit the possibilities and requirements of the discrete, sequential and recursive processes of digital computation. The hydraulics that is reformulated to suit digital machine processes in this way is called *computational hydraulics.*

Computational hydraulics provides formal, mathematical support and guidance for the various techniques used to develop mathematical models of the aquatic environment. By embedding these computational hydraulics models into more general information systems, a class of what is nowadays referred to as Hydroinformatics System emerged. A hydroinformatics system is generally defined as (Abbott, 1991) an electronic knowledge encapsulator that models (part of) the real world and can be used for the simulation and analysis of physical, chemical and biological processes in water, for a better management of the aquatic environment. Therefore, the development of mathematical models which adequately represent our current image of reality is a very important task.

Abbott (1997) gave a general definition of a mathematical model as a collection of (indicative) signs that serve as an (expressive) sign. In this respect, mathematical models of the *aquatic* environment, which are the subject of this study, are representations of knowledge about the physical system and the corresponding hydraulic processes in a form of mathematical relationships expressing known hydraulic principles.

To further explain the relationship between the different forms of representation, the general modelling problem of a physical system is schematised graphically in Figure 1.1.

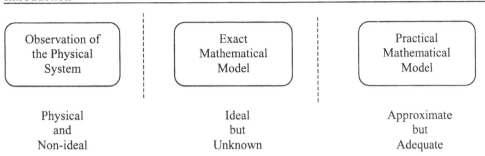

Figure 1.1. A schematisation of different levels of representation of physical systems.

The figure shows three boxes, each representing a particular level of description of a given physical system (e.g., a water body). The left-most of these boxes represents our phenomenological view of the physical system itself, subjected to controllable and uncontrollable influence from its surrounding environment. The middle box represents an ideal, unrealisable, *exact mathematical description* of this physical system. In practice, such a perfect mathematical model is not attainable for either the system or its environment, because almost all mathematical models of real physical systems (i.e. all those other than the simplest celestial motions) involve some degree of approximation. That is, in the process of developing mathematical models, it is necessary to make certain assumptions based upon the fact that either certain effects are believed to be negligible or because more detailed description are simply not available. Thus, *practical mathematical models*, as represented in the right-most box in Figure 1.1. are necessarily approximate since they must be of a manageable complexity and they can only be based on the information available to the model developer. Conversely, in order to be useful, these models must still provide adequate approximations, capturing the most important phenomena exhibited by the physical system. If the model's accuracy is insufficient, the subsequent steps of analysis, prediction or control, which depend on the simulation model, cannot be carried out successfully. There is an obvious trade-off between the required accuracy of the model and its complexity. If the model is too simple, it cannot serve its purpose; on the other hand, the model should not be too complex if it is to be useful in practice. Although on the one hand basic scientific principles can be followed, a great deal of experience and creativity is also involved in developing or constructing practical mathematical models. Therefore, modelling can often be seen as an art as much as a science (Mynett, 1999).

1.1   Current Practices of Computational Hydraulic Modelling

Traditionally, the necessary prerequisites for modelling, within the present context, are a thorough understanding of the system's nature and behaviour, and a suitable mathematical treatment that leads to a usable model. A mathematical model of a hydraulic system can, for the most part, be derived in a deductive manner using those laws of physics which describe the balances of force, mass, momentum and energy. The behaviour of a more general aquatic system can be simulated to some (possibly insufficient) degree of accuracy in terms of other kinds of mathematical representations involving the most important biological and chemical processes occurring in the real natural system. However, a complete understanding of the underlying mechanisms in this latter situation is virtually impossible while even in hydraulics only simplified expressions are normally feasible. Moreover, for distributed parameter systems, these models are usually simplified by exploiting geometric symmetries, neglecting certain spatial non-homogeneities or assuming simplified idealisations. The final result of this development is often a collection of non-linear partial differential, or integral equations that may be represented in *state-space* form, coupled through various constitutive relations and completed with the appropriate initial and boundary conditions. Such a description is said to be *generic* to the class of physical system concerned.

The next stage of the problem is concerned with transforming this *point* or *interval* representation into a representation that is distributed over the entire domain of the solution at all times, such as is for example realised by the process of integrating a partial differential equation. The difficulties experienced in integrating over complicated domains has led to the widespread and now almost universal use of numerical methods in which point and integral descriptions are extended to finite spatial descriptions that are maintained over finite time intervals, thus providing those solution procedures of finite cardinality which are essential for numerical simulation. In doing so, the differential terms in the governing equation are substituted with finite difference, finite element, finite volume or other such terms. The governing equation with the new terms can then be applied simultaneously at all grid points within the model boundary and solved numerically so as to simulate the variations with time in the values of the different flow variables describing the system.

1.2   Problems Associated with the Current Practice

Even though mathematical models are nowadays very widely used in practice, they have a number of potential problems associated with the way they have been re-formulated to be solvable on digital machines. For instance, one approximate translation of the governing equations from the continuum to the discrete form can be obtained by truncating higher order terms in the Taylor series expansion of the differential term, and this usually results in a truncation error. Moreover due to the discretisation used in the numerical model, the variables representing the system's state are computed only at the specified grid points. As a result, depending on the size of the grid spacing, there will be small-scale effects which cannot be considered during the computation. All of this will further increase the discrepancy between the values of the state variables that are used to describe and measure the real physical system and those obtained from the numerical model representing this system.

However, through the many innovations in process identification, parameter estimation, data assimilation and other such developments, numerical-hydraulic models have been used as acceptably accurate and reliable tools for the analysis, design and management of a wide range of water-based assets. But, at the same time, these models have often become very extended. Coastal and dam break models with hundreds of thousands, or even of the order of a million grid points, or urban drainage system models with a similar number of nodes, are becoming increasingly common, and these models make heavy demands on computing capacity and time. Even a quite powerful workstation, which has the greatest computing power that is normally available in practical applications, may be employed for many minutes, and in certain cases for some hours, just to make a single simulation covering, say, a few prototype hours, or at most a few days. Although these times may be acceptable in many situations, such as in design, planning and many management operations, there are a number of applications for which such computing times are unacceptable. The most significant example of these at the present time is that of the real-time control of such assets as river systems or urban drainage systems, where only a relatively short time is often available between a forecast of expected inflow and the need to set counter-measures in action. For example, even with present combinations of long-range weather forecasting, on-line satellite observations and weather radar, a rain event can normally only be predicted with any practically useful accuracy within a horizon of a few hours, while the manoeuvring of an urban drainage system as a means of partially absorbing the resulting heavy flows may also require a similar time. This leaves only a relatively short period available for the elaboration of the necessary optimal control schedules for the pumps, weirs, siphons and other such equipment (see Lobbrecht and Solomatine, 1999).

It is obvious that connecting complex models to a Decision Support Systems (DSS) online, will not meet this criterion. Even if one takes the anticipated increase in processing speed of computers into account, an online connection of mathematical models will still require too much computational time. In addition, ongoing research continues to result in ever more complex physically-based mathematical models, demanding ever more computational time. Experience has shown that the two (the increase in processing speed of computers and the increase in complexity of physically based mathematical models) continue to balance each other, such that whatever becomes the 'average' complex model will always require approximately the same amount of computational time as is presently the case.

A number of reasons could be attributed for the relatively long computational time required for physically based mathematical models. For instance, the main reason for longer computational times during hydrodynamic computations is the very large number of outputs which are computed (in space and time) in spite of the fact that many of these outputs of the model are only intermediate values that may not all be relevant to a particular problem. Sometimes, detailed knowledge of the spatial distribution of a particular state variable may not be necessary; instead, only output at a few particular points, or the average over an area may be required. Modellers are also often not interested in every time scale of the model output; instead, they may only be interested in long-term changes, having no concern for fluctuations on a smaller time scale. This is generally the case with, for example, long-shore sediment transport problems in Coastal Engineering. Regarding cross-shore transport, however, short term effects are often more relevant, since most morphological damage is incurred in a matter of hours during intense storms. In other situations one may be interested only in tidally averaged values, to the extent that these are realistically acceptable.

Given this continues increase in the computational-time requirements of numerical models (usually further exacerbated in the case of real-time control by the need to run a considerable number of simulations when making-up the optimal control strategies) the need arises for new methods that can reduce the time needed to simulate the impact of natural events, such as rainfall, and a number of possible human interventions on hydraulics systems. These difficulties have led to the notion of moving over to a new modelling paradigm that would pass better into current practices and controlling possibilities (Abbott, 1997a). One essential component in this development is that of allowing models to *construct themselves* more or less automatically by learning from all existing numerical-hydraulic models and all available field measurements, in a form that will allow a much more rapid response in terms of outputs to any of a wide range of given inputs.

1.3   Model Induction from Data: an Alternative Approach

Ever since the advent of the von Neuman digital machine about fifty years ago, many researchers have explored the possibility of using machines to perform not merely numerical computations, but also as an aid to performing more human-like tasks, such as learning new concepts, solving new kinds of problems, and so on. As a result there has been an explosive growth of methods in recent years for learning (or estimating dependency) from data, where data refers to the known samples that are combinations of inputs and corresponding outputs of a given physical system. This is not surprising, given the proliferation of low-cost computers (suitable for implementing such methods in the form of software), low-cost sensors and database technologies (for registering, collecting and storing data), and the training of more highly computer-literate application experts (Cherkassky and Mulier, 1998). If such a dependency is discovered, it can be used to predict (or effectively *deduce*) the future physical system's outputs from the known input values.

Already for some time now, it is generally agreed that in order to apply the enormous power of current developments in information and communications technology to the problems of sustainable integrated water resources and environmental management, research cannot be restricted only to the fields of hydraulics and hydrology, but should also be extended to the development and application of modern techniques such as those within the field of artificial intelligence. Recent research in the water resources sector directed towards this modern computational paradigm shows an immense promise to provide innovative solutions to hydraulic and hydrologic problems (Minns and Babovic, 1996). This research work has extended previous work in this area trying to explore some of these relatively new computational paradigms and their possible applications in the area of hydraulic and hydrologic modelling.

The main subject that is addressed in this thesis is, therefore, that of model induction from data for the simulation of hydrodynamic processes in the aquatic environment. Model induction is, in this context, the task of inferring a mathematical description or a mathematical model from a series of measurement on a system. Simulation, on the other hand, is the process of conducting experiments with mathematical models that describe the behaviour of a real physical system on a computer, usually over extended periods of time, with the aim of its long term description, usually for planning, designing and decision-making. Madala and Ivakhnenko (1994) discussed the deductive and inductive approach to mathematical modelling as follows:

> *Deduction* is the application of a general law to many practical problems, while *induction* is the synthesis of a general law from many particular observations... Theorems are proven on the basis of axiomatic theory; thus, scientific ways of model building are conceptualised to be deductive. Other ways of thinking are often referred to as 'non proven' or 'not scientific' or simply 'heuristic' or 'rule of thumb.' However, both ways are equally heuristic, and constrained. The main heuristic feature of the *deductive* approach is the use of an axiom which is based on a *priori* accepted information, whereas the main heuristic for the *inductive* approach is its choice of the external criteria. The choices of either axiomatic or external criteria depend on the experts....

In order to visualise the long-term prospect of the two approaches described in the previous paragraph, it is possible to distinguish the three principal stages of the development of the process of simulation, as shown in Figure 1.2:

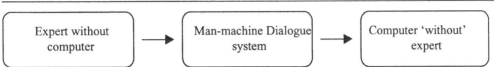

Figure 1.2. Principal stages in the development of simulation and prediction [adapted from Madala and Ivakhnenko, 1994].

In the first stage, predictions are realised in the form of large volumes of data tables compiled on the basis of the reasoning of 'working teams of experts' who basically follow certain 'rules of thumb' which they have established from their long years of practical experience. This approach could still be found in practice, although it is nowadays being replaced for the most part by the second stage of simulation and prediction method which involves the use of both the expert and the computer. The participation of an expert is, in this case, limited to supplying the proper algorithms used in building up the numerical models and the criteria for the choice of the best models with an optimal level of complexity. The process of decision making in the case of incompatible outcomes is resolved by introducing multi-objective criteria. The third stage, which is 'computing without experts' by using what are often called 'artificial intelligence systems', represents systems which are ultimately composed of 'intelligent agents' with no human being actually present (Babovic, 1996). The decisions in artificial intelligence systems are made on the basis of general requests (criteria) of the human user expressed in an abstract metalanguage. Model induction from data is anticipated to be one important component in realising this third stage in the evolution of the development of simulation models of physical systems.

While modelling any physical system (aquatic or otherwise), it is common practice to introduce dependent and independent variables, and one of the important tasks is to know which of the independent variables activate a particular dependent variable. A large number of general methods for doing this are available in the mathematical and statistical literature. Popular among them is applying correlation and regression analysis. Clearly, general methods such as regression analysis are insufficient to represent complex problem-solving skills, but they can form a basis for present-day more advanced methods. In recent years, thanks to the proliferation of low-cost computers, sensors and database technology, there has been a fast growth of methods for inducing models from data. This interest in estimating dependencies from data has resulted in the development of a variety of methods such as artificial neural networks (ANNs), support vector machines (SVMs), evolutionary algorithms (EA) and fuzzy logic. Connectionist models such as artificial neural networks constitute techniques by which learning from data is made possible by describing input-output processes in terms of activation patterns defined over nodes (units) in a highly interconnected network. Information is passed through the units and each individual unit typically will play a role in the representation of multiple pieces of information. The representation of knowledge is thus distributed and is defined by the strength of connections between processing units. In this sense 'the knowledge is in the connections', as connectionist theorists like to put it, rather than in static and monolithic representations of concepts. Learning, when viewed from within this framework, consists of the adjustment of connection strengths between units and the emergence of higher-order structures from more elementary components.

Similarly, support vector machine can also be used for model induction from data. However, SVM is a relatively new machin learning paradigm that is firmly based on the *theory of statistical learning* (Vapnik, 1995). An interesting property of this approach is that it is an approximate implementation of a structural risk minimisation (SRM) induction principle that aims at minimising a bound on the generalisation error of a model, rather than only minimising the mean square error over the data set. Fuzzy models also allow describing complex input-output relationships of a particular problem domain by using the concept of fuzzy sets and fuzzy logic (Zadeh, 1973). They describe logical relationship between linguistic variables by means of if … then … rules, and in this way fuzzy models integrate the logical processing of information with attractive mathematical properties of general function approximation. Therefore, similar to neural networks, fuzzy models are general function approximators which can deal with non-linear and dynamic system quite accurately (Wang, 1992). Fuzzy models are constructed using numerical data and the extracted rules can even provide an interpretation of the systems behaviour.

On the other hand, some researchers have taken the more global view of problem solving as a process of a search through a state space: a problem is defined by an initial state, one or more goal states that have to be reached and, a set of operators that can transform one state into another together with constraints that an acceptable solution must meet (Holland et al., 1986). Then some types of problem-solving techniques are used for selecting an appropriate sequence of operators that will succeed in transforming the initial state into a goal state through a series of steps. A selection approach is taken for classifying the system. This is based on an attempt to impose rules, such as those of 'survival of the fittest', on an ensemble of simple procedures. This ensemble selection process is further enhanced by introducing criterion rules which implement processes of genetic cross-over and mutation on the production of the population. Thus productions that survive a process of selection are not only applied but also used as 'parents' in the synthesis of new populations. Here an external agent is required to play a role in laying out the basic architectures of those productions upon which both selective and genetic operations are performed. These classification systems do not require any *a priori* knowledge of the categories to be identified: the knowledge is very much implicit in the structure of the classifying systems. The concepts of 'natural selection' and 'genetic evolution' are viewed as possible approaches to normal levels of implementation of rules and representations in information processing models.

1.4   Model Structure Selection

Several researchers are experimenting to predict the behaviour of various complex systems in a wide range of physical situations by constructing models from data corresponding to certain variables and using a variety of advanced model induction techniques. An important aspect of using any one of the induction approaches in modelling physical systems is the process of sifting through various sets of models whose complexity is gradually increased, and of testing them for their accuracy. The resulting mathematical models must then be able to extrapolate the behaviour of the systems in the space (x, y, z) co-ordinates, as well as to predict in time (t). The problem goal may be interpolation, extrapolation or prediction and the model that is selected must be the one that provides output that gives the best agreement with the recorded behaviour of the system. In general, the necessary steps in identifying a model of a physical system can be schematised as shown in Figure 1.3.

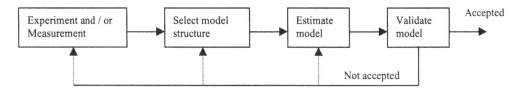

Figure 1.3. Steps in identifying a model of a dynamic system [adapted from Norgaard et al., 2000].

For a model to be accepted, its complexity should be such that it generalises well or, in other words, the model has to perform very well on a validation data set, this being a data set that has never been used during the process of model estimation. If the model is very simple, it may not have the *capacity* to reproduce the training data set even when the best possible set of model parameters are found. On the other hand, if the model is very complex with a very high *capacity*, it will try to model small perturbations which are usually found in the training data (especially if drawn from real world), and it will *over-fit*. The performance of such a model on validation data is often poor, and sometimes very poor, which means that it cannot generalise adequately. Therefore, the choice of the appropriate model structure requires a procedure which make a trade-of between the models *capacity* and its ability to generalise over the class of problem to be modelled.

1.5   Outline of the Thesis

Obviously, this thesis cannot possibly contain a complete exposition of model induction from data. The aim is therefore to elaborate on the main concepts of model induction from data (obtained from field measurements or extracted from simulations using numerical models), to identify the appropriate model induction techniques and to explain how the models so induced can be used as effective simulation tools for some particular situations.

Chapter 2 of this thesis introduces some currently popular artificial neural network architectures such as multi-layer perceptrons and recurrent and dynamic networks, of the kind that have been employed in most of the rest of this study. Chapter 3 highlights yet another model induction techniques, namely the Support Vector Machine, which is emerging as a generalisation of artificial neural networks with the potential to provide more robust behaviour than those obtained from standard ANN techniques. This is demonstrated in two case studies on hydrologic and hydrodynamic modelling problems.

Chapter 4 describes artificial neural network as domain knowledge encapsulator, illustrated by the generation of wave equations from hydraulic data. The emulation of numerical-hydraulic models using artificial neural networks is also discussed. Chapter 5 addresses the question of how (generalised) artificial neural networks can be used as numerical operators on non structured grids for the simulation of hydrodynamic processes in a two-dimensional flow system. This is demonstrated by applying the technique to the simulation of tidal flows in a specific geographic area, namely the Donegal Bay in Ireland. In Chapter 6, a methodology is derived for developing generic hydrodynamic models using artificial neural networks with schemes that can be applied over arbitrary bathymetries with variable distance and time steps. Chapter 7 provides a summary and some conclusions of the thesis, together with some recommendation.

Chapter 2

Artificial Neural Networks as Model Induction Techniques

Based on a conceptual foundation, the process of induction from data is used to build up a model of any given phenomenon or physical system from which it is hoped to simulate the system's behaviour and deduce responses of the system to changes in the environment that have yet to be observed. In recent years, more emphasis is being given to research directed to identifying methods by which these models can be induced directly from the measured data with little or no prior knowledge of the behaviour and the governing laws of the physical system under consideration. These model induction methods use very general meta-languages, rather than a language of detailed instructions. The prerequisite in obtaining a predictive model through the inductive approach is that the quantitative model built from a given set of observations should provide the same result as the model built up from other sets of observations taken at different times and places. In other words, the most important criterion for any model induction technique which uses a limited amount of data is its ability to find a model that provides the widest coverage. As introduced in the first chapter, the capacity of an induction algorithm to find a satisfactory model of a given physical system based on a limited (and in general a relatively) small number of training examples is referred to as its power of generalisation. The power of generalisation is the ability of the model to predict correctly the output of a wide range of samples that have not been encountered during training. The need to address this and other similar problems has resulted in a considerable attention being given lately in research literature to the use of *machine learning* techniques for actually building (or inducing) more generic models automatically from data.

In general, a machine learning method is an algorithm that estimates (induces) a hitherto unknown mapping (or dependency) between a physical system's inputs and its outputs from the available data. However, even though available research literature is concerned with formal description learning methods, there is an equally important informal part of any practical learning system (Cherkassky and Mulier, 1998). This part involves such practical issues as the selection of the input and output variables, the data representation, and the means available for incorporating prior domain knowledge into the designing of the learning system. This informal part is often as critical to the overall success of the learning system as is the design of the learning machine itself. Indeed, if the wrong (uninformative) input variables are used in modelling, then no learning method can provide an accurate prediction. According to Gillies (Gillies, 1996), one has to give attention to the following three features in the process of iterating a basic inductive rule of inference to produce the final result:

(i)   the existence of inductive rules of inference,
(ii)   the role of background knowledge as well as data in these rules and
(iii)   the role of testing and falsification.

Methods for estimating dependencies from data have been traditionally explored in such diverse fields as statistics (multivariate regression and classification), software engineering (pattern recognition), and computer science (artificial intelligence and machine learning). Recent interest in facilitating learning from data has resulted in the development of such biologically-motivated methodologies as artificial neural networks, evolutionary algorithms, fuzzy systems, and indeed several others, such as support vector machines. Since most of the research work presented in this thesis focuses on the use of artificial neural networks as a model induction technique in the fields of hydraulics and hydrology, the rest of this chapter is given over to a brief introduction to this technique.

2.1   Artificial Neural Networks

Historically, computing has been dominated by the concept of programmed computing, in which procedural algorithms are designed and subsequently implemented using whatever happens to be the currently dominant architecture. Since the late 1940s, the dominant architecture has been that of the von Neumann machine in which operands and operators have the same representations, as sequences of binary strings. However, an alternative viewpoint towards programmed computers is needed when one considers the *computing* that is carried out by biological systems. For example, processes of *computation* in the human brain obviously differ greatly from those of the von Neumann paradigm in that, first, the computation is massively distributed and parallel and, second, learning replaces to a great extent formal program development. The development of this biologically-motivated computing paradigm of artificial neural networks (ANNs) began approximately half a century ago (McCulloch and Pitts, 1943). Some ANNs are indeed models of biological neural networks while others are not. However, historically, much of the inspiration for the field of ANNs came from the desire to produce artificial systems capable of sophisticated, perhaps *intelligent*, computations similar to those that the human brain routinely performs, and thereby possibly to enhance our understanding of the human brain. Within the last decade the study of artificial neural networks has experienced a huge resurgence due to the development of more sophisticated algorithms and the emergence of powerful computational tools (ASCE, 2000a). Extensive research has also been devoted to investigate the potential of ANNs as computational tools that acquire, represent, and compute a mapping from one multivariate input space to another.

Artificial neural networks can be defined generally as *flexible mathematical structures that are capable of identifying complex and commonly non-linear relationships between input and output data sets*. A neural network consists of a large number of simple processing elements that are called either neurons, units, or nodes (hereafter, these basic building blocks will be described as *neurons*). Each neuron is then connected to other neurons by means of direct communication links, each being associated with a weight that represents information being used by the net in its effort to solve a problem. The processing of each neuron is broken into two steps (see Fig. 2.1.), that is, the weighted sum of the inputs is taken, and is followed by the application of the activation function. For example, consider a neuron that receives inputs from the input layer. The net input, o_in, to this neuron is the sum of the weighted signals from the input neurons (that is : $o\_in = w_{1i}i_1 + w_{2}i_2 + w_{3}i_3$ ... $w_{n}i_n$). The activation y of this neuron is then given by some function of its net input, $o = f$ (o_in). Essentially, the activation function f can take many forms, but most often it is monotonic. The most popular activation functions, and the ones mostly used in this study, are shown in Figure 2.2.

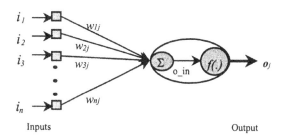

Figure 2.1. A schematisation of an artificial neuron at a node j.

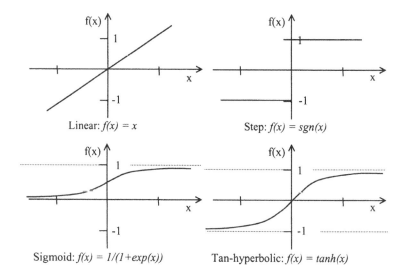

Figure 2.2. Different types of activation function.

A neural network can be in general characterised by its architecture, which is represented by the pattern of connections between the nodes, its method of determining the connection weights, and the activation functions that it employs (Fausett, 1994). As a result, ANNs do not constitute one network, but a diverse family of networks whereby the functionality of each type of network is determined by the network topology, the individual neural characteristics, and the learning or training strategy employed. One way of classifying neural networks is by the number of layers: single layer (Hopfield net), multilayer (most backpropagation networks), etc. ANNs can also be categorised based on the direction of information flow. In a feed forward network, the nodes are generally arranged in layers and information passes from the input to the output layer. On the other hand, in a recurrent network, information flows through the nodes in both directions, from the input to the output layer and vice versa. This is generally achieved by recycling previous network outputs as current inputs, thus allowing some degree of feedback. Still another way of classifying ANNs is by distinguishing between networks with *supervised learning,* where the networks are provided with training patterns of input-output pairs from which they try to set optimum sets of parameters (weights), and *unsupervised (or competitive) learning* where the networks extract information from input patterns alone, without the need for a desired response or output. However, there it is not the intention to discuss about the various possible categories further in this work. Instead, some typical and more widely used types of network architectures will be reviewed in the following sections.

2.2   Multi-Layer Perceptron Network

A typical neural network consists of a number of nodes that are organised according to a particular arrangement. Multi-layer perceptrons (MLPs; see Fig. 2.3.), which constitute probably the most widely used network architecture, are composed of a hierarchy of processing units organised in a series of two or more mutually exclusive sets of neurones or layers. The first, or input, layer serves as a holding site for the input applied to the network. This consists of all quantities that can influence the output. The input layer is thus transparent and is a means of providing information to the network. The last, or output, layer is the place at which the overall mapping of the network input is made available, and thus represent model output. Between these two layers lie one or more layers of *hidden* units. The information flow in the network is restricted to a flow, layer by layer, from the input to the output (see Fig. 2.3.). In this figure, $i = [i_1, i_2, \ldots, i_n]^t$ is a system input vector composed of a number of causal variables that influence system behaviour, and $o = [o_1, o_2, \ldots, o_m]^t$ is the system output vector composed of a number of resulting variables that represent the system behaviour. The unidirectional nature of the information flow places

MLPs amongst what are usually called feed-forward networks. Each layer, based on its input and connection weights, computes an output vector and propagates this information to the succeeding layer. The mathematical characterisation of an MLP network is that of a composite application of functions. Each of these functions applies to a particular layer, but may further be specific to individual units in the layer. The overall mapping is thus characterised by a composite function relating the network's input to its output. This can be described using the arrow notation of category theory (Abbott and Dibike, 1998) as will be explained below.

Since the nodes in the input layer do not affect the inputs but just pass these to the next layer, the transfer function f_1 on the input layer can be considered as an identity function, that is:

$$f_1(i) = i \tag{2.1}$$

Then the next function f_2 provides the summation of these inputs multiplied by the corresponding connection weights and it can be described as follows:

$$f_2(i) = \sum_{j=1}^{n} w_j i_j = a \tag{2.2}$$

The output a of this function f_2 is once again an input for the transfer function in the output layer. This function f_3 can take any one of the commonly used forms, such as sigmoid, Gausian or even linear, and gives o as its output. Take, for example, the sigmoid function:

$$f_3(a) = \frac{1}{1 + e^{-a}} = o \tag{2.3}$$

Then, using the arrow notation of category theory, one can describe the above functional relationship as follows:

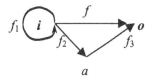

then by composing the above functions together :

$$i \xrightarrow{\ f\ } o$$

This can be described as:

$$f(i) = f_3 \circ f_2 \circ f_1 = f_3\big(f_2\big(f_1(i)\big)\big) = o \tag{2.4}$$

This composition of functions can be extended for the case of multi-layer perceptrons with one hidden layer as follows:

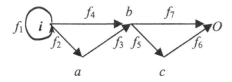

This can be further simplified as :

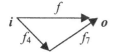

where f_1 , $f_2 = f_5$, and $f_3 = f_6$ are as explained before. However, f_5 and f_6 are here the corresponding functions applied between the middle and output layer.

Therefore one can write this ANN, once again

$$i \xrightarrow{\ f\ } o$$

where $f(i) = f_7 \circ f_4 = f_6 \circ f_5 \circ f_3 \circ f_2 \circ f_1 = f_6\big(f_5\big(f_3\big(f_2\big(f_1(i)\big)\big)\big)\big) = o \tag{2.5}$

In general, an MLP network with one hidden layer has been shown to provide a universal function approximator (Hornik et al., 1989). However, the design of an MLP network for a specific application may involve many other issues, most of which require problem-dependent solutions.

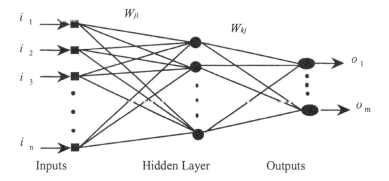

W_{ji}

W_{kj}

Inputs   Hidden Layer   Outputs

Figure 2.3. Schematisation of a multi-layered, feed forward neural network (MLP).

2.2.1 Learning in MLP networks (training)

Although each network architecture is associated with particular mode of learning and so different operational principles, the characterisation of a system by the generic term *neural network* implies simply an ability to learn in a networked fashion. Let us consider learning (or training) in ANNs, mathematically, as an approximation of the actual multi-variable function $g(i) = t$ by another function $g'(i, w)$, where $i = [i_1, i_2, \ldots, i_n]^t$ is the input vector, $t = [t_1, t_2, \ldots, t_m]^t$ is the corresponding output (target) vector and $w = [w_1, w_2, \ldots, w_{mn}]^t$ is a parameter (weight) vector. The learning task is then to find the weight w that provides the best possible approximation of $g(i)$ in some predefined sense based on the set of training examples i and t. Given this training set of input-output data, the most common learning rule for multi-layer perceptrons is that of back-propagation based upon what is usually called a *generalised delta rule* (GDR). The GDR is a learning rule that uses a method of *gradient descent*, as will be described later, to achieve training or learning by error correction, and a neural network with this type of learning algorithm is usually referred to as a back propagation network (McClelland and Rumelhart, 1988). Each input pattern of the training data set is passed through the network from the input layer to the output layer. The network output is compared with the desired target output and an error is computed. This error is then propagated backward through the network to each node, and correspondingly the connection weights are adjusted based on an equation of the general form:

$$w(n+1) = w(n) + (\Delta w) \qquad\qquad (2.6)$$

where $w(n)$ specifies the connection weights obtained during the current iteration (n) and (Δw) the required change in weight necessary to calculate the weights for the next, $(n+1)^{th}$ iteration.

Let us first consider a network with no hidden layer and let i represent an input pattern, o the corresponding output pattern, w the network vector of weights and t the desired (target) system output. The training set consists of ordered pairs of vectors and is denoted by $H = \{(i^p, t^p)\}$ $p = $ 1, 2, ..., n, where, for the pth input-output pair, i^p_i is the ith input value. Similarly o^p_j and t^p_j are the jth elements of o^p and t^p, respectively and w_{ij} represents the weight of the connection from i_i to o_j. The output error vector for the pth pattern pair can then be defined as:

$$e^p = t^p - o^p \tag{2.7}$$

A scalar measure of the output error based on the pth training sample is denoted E^p and is conventionally defined as:

$$E^p = \frac{1}{2}\sum_j \left(t^p_j - o^p_j\right)^2 \tag{2.8}$$

Over the whole training set H, the total error is then

$$E = \sum_{p=1}^{n} E^p \tag{2.9}$$

The basic idea is to compute $\partial E^p / \partial w$ and use this quantity to adjust w. This can be written as:

$$\frac{\partial E^p}{\partial w_{ji}} = \frac{\partial E^p}{\partial o^p_j} \frac{\partial o^p_j}{\partial net^p_j} \frac{\partial net^p_j}{\partial w_{ji}} \tag{2.10}$$

Where $net^p_j = \sum_i w_{ji} i^p_i$, $o^p_j = f(net_j)$ and f is the activation function

Then :

$$\frac{\partial E^p}{\partial o_j^p} = -(t_j^p - o_j^p) = -e_j^p \tag{2.11}$$

$$\frac{\partial o_j^p}{\partial net_j^p} = f'_j(net_j^p) \qquad\qquad \text{and} \tag{2.12}$$

$$\frac{\partial net_j^p}{\partial w_{ji}} = i_i^p \tag{2.13}$$

Therefore,

$$\frac{\partial E^p}{\partial w_{ji}} = -(t_j^p - o_j^p)f'_j(net_j^p)i_i^p \tag{2.14}$$

$$= -(\delta_j^p)i_i^p$$

where $\delta_j^p = (t_j^p - o_j^p)f'_j(net_j^p)$

By choosing f to be a *sigmoid* function i.e. a specific non-decreasing and differentiable function defined by:

$$o_j^p = f(net_j^p) = \frac{1}{1 + e^{-net_j^p}} \tag{2.15}$$

we find that:

$$f'(net_j^p) = \frac{e^{-net_j^p}}{(1 + e^{-net_j^p})^2} = f(net_j^p)(1 - f(net_j^p)) \qquad \text{or}$$

$$f'(net_j^p) = o_j^p(1 - o_j^p) \tag{2.16}$$

An intuitively reasonable iterative weight correction procedure using the pth training sample can then be derived, using

$$\Delta^p w_{ji} = -\varepsilon\left(\frac{\partial E^p}{\partial w_{ji}}\right) \tag{2.17}$$

$$= \varepsilon\delta_j^p \, i_i^p$$

where ε is a positive constant, refered to as the learning rate. This is the basic principle of gradient descent.

To generalise the results to networks with units on the hidden layers, the general representation for the ith input to neuron j, denoted $\tilde{o}_i^p$, is defined as

$$\tilde{o}_i^p = \begin{cases} o_i^p & \text{if this input is the output of another neuron} \\ i_i & \text{if this input is a direct input to the network} \end{cases}$$

$$\frac{\partial net_j^p}{\partial w_{ji}} = \tilde{o}_i^p \qquad\qquad (2.18)$$

Therefore, in the case of output units, the procedure used to modify the weights uses the expression

$$\Delta^p w_{ji} = \varepsilon \delta_i^p \tilde{o}_i^p \qquad\qquad (2.19)$$

With some modifications needed in order to take account of the influence on the other n outputs units, the weights of the hidden unit u_k are modified as

$$\Delta^p w_{ki} = \varepsilon \delta_k^p \tilde{o}_i^p \qquad\qquad (2.20)$$

where $\delta_k^p = f'_k(net_k^p)\sum_n \delta_n^p w_{nk}$, with the δ_n^p being taken from the next $(k+1)^{th}$ layer.

The value of the learning rate ε depends on the characteristics of the *error surface* which represents the variability of the error value as a function of the weight vector. If the surface changes rapidly, a smaller learning rate is desirable; on the other hand, if the surface is relatively smooth, a larger learning rate will accelerate the convergence. This rationale, however, is based on a knowledge of the shape of the error surface, which is rarely available. Often, the scaling parameter ε is adjusted as a function of the level of the iteration. Moreover, a so-called *momentum* is usually employed in training formulations as a way of introducing some inertia into the procedure and so damp-out oscillation in the system. This helps the system escape local error function minima in the training process by making the system less sensitive to local changes (Schalkoff, 1997). In order to add a *momentum* to the weight update at the *(n+1)*th iteration, the correction Δw_{ji} is modified as follows:

$$\Delta^p w_{ji}(n+1) = \underbrace{\varepsilon \delta_j^p(n+1)\tilde{o}_i^p(n+1)}_{\text{as before}} + \alpha\Delta^p w_{ji}(n) \qquad\qquad (2.21)$$

Usually the weight correction procedure is based on a so-called *epoch error*, instead of on the individual input patterns. This is to say that the procedure is repeated for each input vector and, at the completion of a pass through the entire data set, all the nodes change their weights based on the accumulated weight changes, as expressed by:

$$\Delta w_{ji} = \sum_{p} \Delta^{p} w_{ji} \qquad (2.22)$$

Despite its popularity, the use of backpropagation learning also introduces some difficulties. The first difficulty is the necessity of providing a prior specification of the network structure. If the size of a network (number of hidden layers and the number of neurons on each hidden layer) is too large, the network can be expected to generalise quite poorly. On the other hand, if it is too small, learning from training samples becomes insufficient to provide an adequate generalisation. Since prior structural information is usually not available, identifying the optimum network structure then usually becomes a matter of trial and error, which can sometimes be a time consuming process. The second difficulty is a so-called *local minima* problem in which the network identifies only one local minimum in its objective function while ignoring other and more significant minima. This can become more and more serious as the network size increases. One of the methods that have been proposed to address the first problem is called *structural learning*. This method introduces various 'pruning algorithms', which remove hidden units and connections which demonstrate only minor contributions to the error function, and the introduction of a form of weight decay by placing a penalty on the associated error criteria. An alternative solution, which helps to avoid spurious local minima, is to take account of second order effects in the gradient. For example, the performance of the back propagation procedure can also be improved by using an approximation of Newton's method, called in this case the Levenburg-Marquardt method. It is claimed that this approximation technique is more powerful than is that of direct gradient descent, but it clearly requires more memory during computation (Demut and Beale, 1994).

Although MLPs are the most common and widely used types of artificial neural networks, there are also other types of network architectures which are useful to induce models from data. Some have similar feed-forward structure like MLPs, while others have dynamic recurrent structures; some use supervised learning for training while others employ unsupervised learning. Brief overviews of some of these neural network architectures are given below.

2.3   The Radial Basis Function (RBF) Network

The most common alternative to the multilayer perceptron network is probably the Radial Basis Function (RBF) network (Norgaard et al., 2000). A radial basis function network has again a feed-forward structure, but with a modified hidden layer which performs a fixed nonlinear transformation. While there are various choices, the mapping function of a radial basis function network, as schematised in Figure 2.4. is mostly built up of Gaussians rather than the sigmoid of the MLP networks. RBF network also has a different learning (training) algorithm (Mason et al., 1996) which is carried out in two phases, first for the hidden layer, and then for the output layer. The hidden layer is self-organising; its parameters depend on the distribution of the inputs, not on the mapping from the input to the output. The output layer, on the other hand, uses a supervised learning (gradient decent or linear regression) in order to set its parameters.

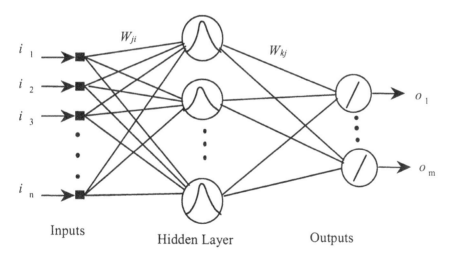

Figure 2.4. Schematisation of radial basis function (RBF) network.

An RBF hidden unit has one parameter associated with each input unit. These parameters w_{ij} are not weights placed on the input; rather they are the co-ordinates in input space of a point that is the centre of the Gaussian function of the hidden units. The output from the hidden layer is a function of the radial distance net_j between the input vector $\boldsymbol{i} = (i_1, i_2, \ldots, i_k)$ and the *radial centre* $\boldsymbol{w_j} = (w_{1j}, w_{2j}, \ldots w_{mj})$ and may be written as :

$$net_j = \| \boldsymbol{i} - \boldsymbol{w}_j \|^2 \quad , \tag{2.23}$$

It is then possible to describe the output for the hidden layer by $o_i = f(net_j)$. There are various choices of $f(net_j)$, and the most popular form is that of Gaussian described as follows (Taylor, 1996):

$$o_j = f(net_j) = e^{-\| \boldsymbol{i} - \boldsymbol{w}_j \|^2 / 2\sigma_j^2} \tag{2.24}$$

where σ is the standard deviation of the Gaussian function

The hidden units thus send signals to the output units that indicate how far the example is from the centre of each activation function. The output units are most often simple linear units and each has a parameter for each hidden unit. The network's outputs are then calculated as:

$$o_k = \sum_{j=1}^{J} w_{kj} o_j \tag{2.25}$$

where w_{kj} is the parameter on the connection from the hidden node j to the output node k and o_j is the output of the hidden node j.

As already introduced, learning – which here consists only of finding the values of the weights - is carried out in two phases, the first applying to the hidden layer and the second to the output layer. The hidden layer is, as stated earlier, self-organising: its parameters depend on the distribution of the inputs, not on the mapping from input to the output. Therefore we can use the so-called k-means or SOFM clustering algorithms to train RBF networks (Hassoun, 1995). The output layer, on the other hand, uses supervised learning (gradient descent or linear regression) to set its parameters. In general, RBF networks that employ clustering for locating hidden unit receptive field centres can achieve a performance comparable with back propagation networks while requiring orders of magnitudes less training time. However, the RBF network usually requires more data to achieve the same accuracy as back propagation networks.

2.4 Recurrent Networks (as species of dynamic networks)

In temporal problems, the measurements from the outer world are no longer an independent set of input samples, but functions of time. A single sensor produces a sequence of measurements that are linked by an order relation, defining the signal time structure. If one changes the order of the samples, one is distorting the time signal, and changing its frequency content, so sample order must be preserved in temporal processing (Principe et al., 2000). To exploit the signal's time structure, the learning machine must have access to the time dimension. Since physical systems are causal, the search is restricted to the past of the signal. Physical structures that store the past of a signal are called *short-term memories*, or simply *memory structures* and it is now possible to incorporate memory structures inside a learning machine. This means that instead of using a *window* over the input data, we create neurons dedicated to storing the history of the input signal.

While feedforward neural networks are popular in many application areas, they lack the feedback connections necessary to provide a dynamic model. There are, however, more recent explorations of neural networks that have feedback connections and are thus inherently dynamic in nature. Dynamic neural networks are topologies designed to include time relationships explicitly in the input-output mappings. The application of feedback enables the networks to acquire *state* representations, which make them suitable for non-linear prediction and modelling (Gautam and Holz, 2000)

There are different ways of introducing 'memory' in order to build a dynamic neural network. For instance, a network can be formulated by replacing the neurons of an MLP with a memory structure, which is sometimes called a *tap delay line*. If the memory structure is incorporated only with the input neurons, then it is called the *focused* time delay neural network (TDNN). It is possible to design an MLP equivalent to the focused TDNN by using multiple samples from the time series instead of a single input and the tap delay line, that is, by placing a rectangular window over the time series and choosing an MLP with as many inputs as samples in the window. However, by bringing the memory inside the learning machine, in the case of the TDNN, the efficiency of the representation of time is expected to improve since the learning machine can receive only the present sample from the external world. The size of the memory layer (the tap delay) depends on the number of past samples that are needed to describe the input characteristics in time and it has to be determined on a case-by-case basis. One interesting feature of the TDNN is that the tap delay line at the input does not have any free parameters; therefore the TDNN can still be trained with a static backpropagation. The *focused* TDNN topology has been successfully used in nonlinear system identification, time series prediction, and temporal pattern recognition (Principe et al., 2000).

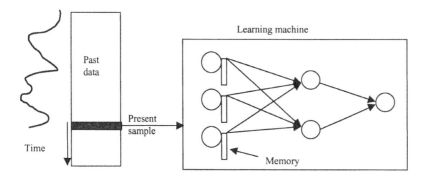

Figure 2.5. Schematisation of the time delay paradigm for processing time series data [adapted from Principe et al., 2000].

The other type of dynamic neural network is composed of networks with recurrent connections among some or all neurons which form closed loops in the network topology. These are known as recurrent networks. If the recurrent network uses fully recurrent connections, then it is often referred to as a Hopfield net (Hopfield, 1982). Patterns stored in this configuration correspond to the stable states of a non-linear system. This device is able to recall, as well as complete, partially specified inputs (Schalkoff, 1997). It may also employ different types of learning methods, such as backpropagation through time (BPTT)(Williams and Zipser, 1995). However, the popular way to recognise (and ultimately to reproduce) sequences has been to use partially recurrent networks. In this case, the network connections are mainly feed-forward, but include a carefully chosen set of feedback connections. The introduction of recurrence allows the network to remember cues from the recent past, while at the same time not appreciably complicating the training. Simple recurrent networks like, Jordan networks (Jordan, 1986) and Elman networks (Elman, 1990), are primarily feed-forward structures with additional feedback loops for implementing the dynamics. The specific group of units which receive feedback signals from the values already obtained at the previous time steps are known as *context units* and they form a *context layer*. Elman's context layer receives input from the hidden layer, while Jordan's context layer receives input from the output. Both the Jordan and Elman nets have fixed feedback parameters and there is no recurrency in the input-output path; therefore, they can be approximately trained with a straightforward backpropagation algorithm.

In most dynamic systems, there is an algebraic relationship between prediction on the one hand and past measurements and external inputs on the other hand. In linear systems the use of the past of the input variable or forcing creates what is commonly called a finite-impulse response (FIR) model structure (Norgaard et al., 2000). Accordingly, a time-delay neural network (TDNN), as shown in Figure 2.5., can be considered as a non-linear extension of a finite-impulse response model structure that is sensitive to temporal relationships in the forcing. On the other hand, the use of the past of the observed output in addition to the forcing, creates what are called auto-regressive models with exogenous input (ARX). However, if, in addition to the external input, past values of the system's output or the prediction error are feed back as input, then the network provides what is called an auto-regressive moving average model with exogenous input (ARMAX) (see Fig. 2.6.).

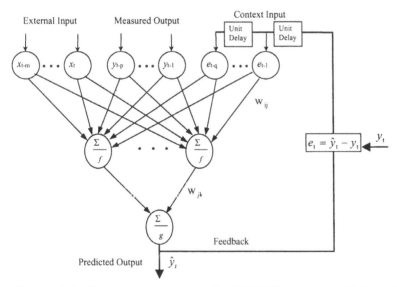

Figure 2.6. Recurrent network of ARMAX structure [Adopted from Drossu and Obradovic, 2000].

In the case of neural nets, these topologies also become NARX and NARMAX model structures respectively, where the prefix 'N' stands for 'non-linear' (Boogaard et al., 1998). In other sections of the present work, this feedback of output to the input layer will be regarded as providing the network with the initial state of the system (for example, the value of the flow variables, such as the water depth in an estuary, at time *t-1*) in order to calculate the present state (that is, in that case the value of the flow variables at the next time, *t*).

Neural network models of the above type can be used to predict either one step ahead in time or further into the future. For a prediction horizon larger than one time step, the prediction can proceed in a direct or in an incremental fashion (Drossu and Obradovic, 2000). In the direct approach, the network is trained to predict directly the *s*-step ahead without predicting any of the intermediate, 1, …, *s*-1, steps. In the incremental approach, the neural network predicts all the intermediate values up to *s* steps ahead by using previously predicted values as inputs when predicting the next value.

One of the difficult problems in the processing of time series is to decide the length of the time window. Normally we do not know the length of time where the information relevant to processing the current signal sample resides. The value of the feedback parameter controls the memory depth in the context units, so that, in principle, its extraction by adaptation from the data may solve the problem. An appealing idea for time processing is to let the system find the memory depth that it needs to represent the past of the input signal. If we utilise the information of the output error to adapt the feedback parameter μ_1, then the system will work with the memory depth that provides the smallest error (such as MSE). This is possible in principle since the feedback parameter μ_1 is related in a continuous way (such as through a decaying exponential) to the units output. In order to adapt the feedback parameter, we will then need to compute the sensitivity of the output to a change in μ_1. The fundamental difference between the adaptation of the weights in static and recurrent networks is that only in the latter do the local gradients depend on the time index. Moreover, the types of optimisation problems are also different in case of dynamic networks since the performance of adaptation is quantified within a time interval, instead of instantaneously. The most common error criterion for this type of dynamic neural networks is provided by a so-called *trajectory learning,* where the cost in error is summed over time from some initial time until the final time of the trajectory. A cost function is therefore obtained over a time interval, and the goal is to adapt the weights to minimize the cost of the error over the time interval.

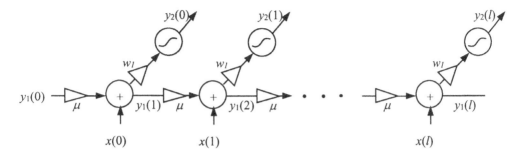

Figure 2.7. Unfolding in time of a simple recurrent network [taken from Principe et al., 2000].

In order to apply the backpropagation procedure for such recurrent networks, we need to adapt the ordered list of dependencies for recurrent topologies. Generally the present value of the activation (also called the state) depends on its previous value and, moreover, the sensitivity of the present state also depends on the previous sensitivity. There is a procedure, called unfolding in time (Principe et al., 2000) that produces a time-to-space mapping that replaces a recurrent network with a much larger feedforward network with repeated coefficients. As long as the network is feedfoward, we can apply the ordered list and each 'time-tick' creates a stage in the unfolding process. Each additional time-tick will thus produce a corresponding network cascaded on the previous one (Fig. 2.7.). Therefore, provided that the recurrent network is in operation for a finite time, we can replace it with a static, feedforward net. This operation is referred to as 'unfolding the network in time'. Probably the most important advantage of proceeding in this way is that, by unfolding the network in time, we are still able to apply the backpropagation procedure to train the recurrent network. This new form of backpropagation is called backpropagation through time (BPTT). The data-flow algorithm of static backpropagation must of course be adapted to cope with the time-varying gradients. Due to the fact that BPTT is not local in time, we have to compute the gradients of all the weights and states over a time interval, even if only a part of the system is recurrent.

It is possible to say that a recurrent system is naturally trained in time since it is a dynamic system. Thus trajectory learning is rather important in many dynamical applications of neural networks. Static networks cannot learn time trajectories directly, since they do not possess a dynamics of their own.

2.5   Self-Organising Neural Networks

An important feature of neural networks is their ability to learn from their environment, and, through learning, to improve their performance in some sense. The networks discussed in the previous sections implemented supervised learning, where a set of targets of interest was provided by an external teacher. The targets could take the form of a desired input-output mapping that the algorithm was required to approximate. However, the purpose of a network for self-organised learning is to discover significant patterns or features in the input data, and to make the discovery without a teacher. In order to do this, the algorithm for the network can be provided with a set of rules of a local nature, which will enable it to learn to compute an input-output mapping with specific desirable properties; the term 'local' means that the change applied to the synaptic weight of a neuron is confined to the immediate neighbourhood of the neuron.

Self-Organizing Feature Maps (SOFM), also known as Kohonen maps, were first introduced by von der Malsburg (1973) and in their present form by Kohonen (1982). The structure of a self-organising network usually consists of an input (source) layer and an output (presentation) layer, with feed forward connections from input to output and lateral connections between neurons in the output layer, as shown in Figure 2.8. The learning process corresponds to repeatedly modifying the synaptic weights of all the connections in the system in response to input (activation) patterns and in accordance with prescribed rules, until a steady configuration develops.

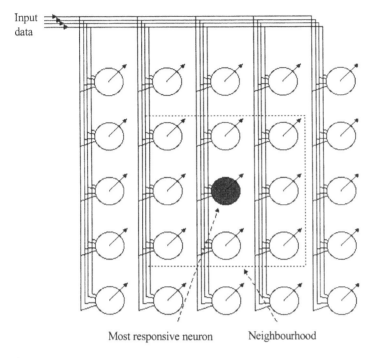

Most responsive neuron   Neighbourhood

Figure 2.8. Architecture of Kohonen's Self Organising Feature Map.

Unlike feed forward and recurrent networks, that are primarily used for approximation and classification, SOFMs are typically used for density estimation or for projecting patterns from high dimensional to low dimentional spaces (ASCE, 2000). In general, a self-organizing network attempts to map a set of input vectors $\mathbf{x}^k$ in $\mathbf{R}^n$ onto an array of units (normally one- or two-dimensional) such that any topological relationships among the $\mathbf{x}^k$ patterns are preserved and are represented by the network in terms of a spatial distribution of unit activities. Kohonen's self-organizing feature map (Kohonen, 1995), which is one of the most prominent of the available topology-preserving models, is used to capture the topology and probability distribution of the input data. This model generally employs an architecture consisting of a two-dimensional structure (array) of linear units, where each unit receives the same input $\mathbf{x}^k \in \mathbf{R}^n$. Each unit in the array is characterised by an n-dimensional weight vector. The ith unit weight vector $\mathbf{w}_i$ is sometimes viewed as a position vector that defines a virtual position for a unit i in $\mathbf{R}^n$. This, in turn, will allow an interpretation of changes in $\mathbf{w}_i$ as a movement of unit i, even though of course no physical movements of units really takes place.

The learning rule for Kohonen's self-organising feature map is defined by:

$$\Delta \mathbf{w} = \rho \Phi\left(r_i, r_{i^*}\right)\left(\mathbf{x}^k - \mathbf{w}_i\right) \qquad \text{for all } i = 1,2,... \qquad (2.26)$$

Where i^* is the index of the winning unit and ρ, $(0 < \rho < 1)$, is a learning rate, which decreases in time.

The winning unit is the node with the weight closest (in the Euclidean distance sense) to the input vector. The neighbourhood function $\Phi(r_i, r_{i^*})$ is very critical for the successful preservation of the topological properties. It is normally symmetric, with values close to 1 for units i close to i^*, and decreases monotonically with the Euclidean distance $\| r_i - r_{i^*} \|$ between unit i and i^* in the array. The following is one possible choice for the neighbourhood function:

$$\phi\left(r, r_{i^*}\right) = e^{-\frac{\| r_i - r_{i^*} \|^2}{2\sigma^2}} \qquad (2.27)$$

where the variance σ^2 controls the width of the neighbourhood.

In general, the two steps in the process which captures the basis for various self-organising algorithms are: first locate the best-matching unit for the input vector $\mathbf{x}$, and then increase matching at this unit and its topological neighbours. These two steps are repeated until a steady configuration develops.

Self-organising neural networks are mostly used in hydrological problems to extract important properties of the input data and to map input data into a 'representation' domain. Other algorithm or networks can then take the training data with the clusters detected through SOFM as input for subsequent supervised learning or classification. For instance, recent studies by Hall et al. (2000) on the application of ANNs on regional flood frequency analysis have shown that catchments may be classified on the basis of their mapped characteristics by the use of supervised learning with a Kohonen self-organising feature map (SOFM), and the parameters of the at-site frequency distributions of annual floods may be related to catchment characteristics through supervised learning with an MLP type of network. Breitscheidel et al. (1998) have also shown the applicability SOFM to classifying storm situations in the past so that it will give experts an indication how the weather (storm situation) could further develop and which impact of the water levels at the Dutch coast it might have.

Further developments in other areas of model induction techniques, such as dimensionally constrained genetic programming (Babovic and Keijzer, 2000) has also shown the potential for modelling and simulation of a wide range of physical processes. For the purpose of the present study, one such relatively new innovation which is considered to be a more general form of artificial neural networks, namely that of Support Vector Machines, will be introduced in the following chapter.

Chapter 3

Model Induction with Support Vector Machines

Artificial neural networks, and in particular multilayer perceptrons and recurrent networks, have gained an immense popularity in recent years as data-driven modelling tools in hydraulics and hydrology since they can be made to perform reasonably well in most cases of practical interest. However, there are also some legitimate concerns on some aspects of ANNs which may still need further investigation. For instance, the network's architecture has to be determined a priori or modified while training by some heuristic which may not necessarily result in an optimal structure of the network (Smola, 1996). A large number of inputs to a network usually results in a more complicated architecture with a greater number of parameters to be optimized. Moreover, the capacity of the networks is controlled only indirectly by such methods as stopping the training early, pruning or weight decay in order to obtain better generalisation. Neural networks can sometimes get stuck in local minima while training, making the network unable to approximate the data set. There are a number of researches directed towards addressing these problems and some have resulted in a more general form of learning systems, called Support Vector Machines (SVMs)

3.1   Introduction to Support Vector Machines

Support Vector Machines are learning systems that have been recently developed upon a framework of statistical learning theory (Vapnik, 1998; Cortes and Vapnik, 1995), and have been successfully applied in a number of applications, ranging from time series prediction (Muller et al., 1997; Mattera and Haykin, 1999), and digital image recognition to identifying land-use patterns (Dibike et al., 2000 // 2001). Due to their many attractive features and their promising performances, SVMs are gaining popularity in the areas of learning from data.

The theory of SVMs is based upon the so-called Structural Risk Minimisation (SRM) induction principle, which has been shown to be superior to the more traditional Empirical Risk Minimisation (ERM) principles employed in many other modelling techniques (Osuna et al., 1997; Gunn, 1998). The architecture of a SVMs does not have to be determined before training and the modelling function may be chosen among a great variety of possible options. Moreover, SVMs are trained by solving a constrained quadratic optimisation problem, which implies that there is a unique optimal solution for each choice of the SVM parameters. This is essentially different from other learning machines, such as standard neural networks trained using backpropagation, and it provides SVMs with a greater ability to find a global optimum and better generalise the data.

In statistical learning theory, the problem of learning is generally viewed as that of choosing, from the given set of functions $f(\mathbf{x}, \alpha)$, $\alpha \in \Lambda$, the one that best approximates a given input-output data set. The selection of the desired function is based on the training set of l independent and identically distributed observations $(\mathbf{x}_1, y_1)$, ... $(\mathbf{x}_l, y_l)$ drawn according to a probability distribution $P(\mathbf{x}, y)$. If one considers the expected value of the error, given by the *risk functional*

$$R(\alpha) = \int \frac{1}{2} |f(\mathbf{x},\alpha) - y| dP(\mathbf{x},y),$$ (3.1)

the objective becomes one of finding the function $f(\mathbf{x}, \alpha_0)$ that minimises the risk functional $R(\alpha)$ in the situation where the only available information is contained in the training set. Since the underlying distribution $P(\mathbf{x},y)$ is unknown, an induction principle is needed in order to infer from the l available training examples a function that minimises the expected error (or *risk*). One such principle is that of Empirical Risk Minimisation (ERM) over a set of possible functions, which is said to provide a hypothesis space. Formally this can be expressed as the objective of minimising the empirical error:

$$R_{emp}(\alpha) = \frac{1}{l}\sum_{i=1}^{l}\frac{1}{2}|f_{\alpha}(\mathbf{x}_i) - y_i| \tag{3.2}$$

with f being restricted to be in a space of functions – or hypothesis space - say S.

An important question is that of how close the empirical error of the solution (or of the device that minimises the empirical error) is to the minimum of the expected error that can be achieved with functions from S. The theory of uniform convergence in probability, developed by Vapnik and Chervonenkis (VC), provides bounds on the deviation of the empirical risk from the expected risk. This theory shows that it is crucial to restrict the class of functions that the learning machine can implement to one with a *capacity* that is suited to the amount of available training data.

For $\alpha \in \Lambda$ and $l > h$, a typical uniform VC bound, which holds with probability 1-η, has the following form:

$$R(\alpha) \le R_{emp}(\alpha) + \sqrt{\frac{h(\log\frac{2l}{h}+1) - \log(\eta/4)}{l}} \tag{3.3}$$

The parameter h is called the *VC (Vapnik-Chervonenkis)-dimension* of a set of functions and is defined as the maximum number of vectors that can be separated into two classes in all 2^h possible ways, using functions of the set. It describes the *capacity* of the set of functions to represent the data set, and in order to minimise the actual risk $R(\alpha)$ one has to minimise the right-hand side of the inequality in Equation (3.3) simultaneously over both terms, and for that one has to make the VC dimension a controlling variable. The VC 'confidence term' in Equation (3.3) depends on the chosen class of functions, whereas the empirical risk depends on the one particular function chosen by the training procedure. Therefore, from these bounds that the theory provides, it is possible to improve the ERM inductive principle by considering a structure of hypothesis spaces $S_1 \subset S_2 \subset ... \subset S_n$, with ordered 'complexity' (i.e. S_{i+1} is 'more complex' than S_i, see Fig. 3.1.). The ERM is performed over each of these spaces, and the choice of the final solution can be made using the aforementioned bounds. This principle of performing an ERM over a structure (sequence) of nested hypothesis spaces is known as Structural Risk Minimisation (SRM)

(Vapnik, 1998). SRM then consists of finding that subset of functions which minimises the bound on the actual risk.

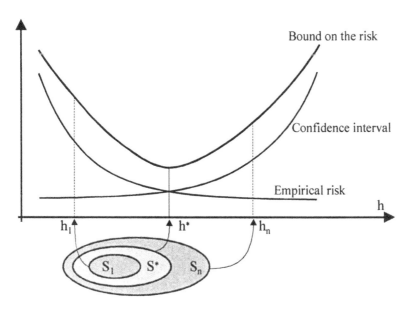

Figure 3.1. The bound on the actual risk is the sum of the empirical risk and the confidence interval [adopted from Vapnik, 1998].

3.2   Support Vector Classification

For the case of a two-class pattern recognition problem, given a set of examples, i.e. pairs of patterns x_i and labels y_i, (x_1, y_1), . . . (x_l, y_l), $\in \mathbf{R}^N \times \{-1, +1\}$, each one of them generated from an unknown probability, and a set of functions $\{f_\alpha : \alpha \in \Lambda\}, f_\alpha : \mathbf{R}^N \rightarrow \{-1, +1\}$, then the task of learning from examples can be reduced to one of learning a function f_α which provides the smallest possible value for the average error committed on independent examples randomly drawn from the same distribution, called the *risk*. This can be formulated as a problem of dividing the set of training vectors belonging to two separate classes with a hyperplane

$$(\mathbf{w}.\mathbf{x}) + b = 0 \quad , \qquad\qquad \mathbf{w} \in \mathbf{R}^n \quad and \quad b \in \mathbf{R} \qquad\qquad (3.4)$$

which is induced from the available examples corresponding to decision functions $f(\mathbf{x}) = \text{sign}(\mathbf{w} . \mathbf{x} + b)$ such that it will work well on *unseen* examples.

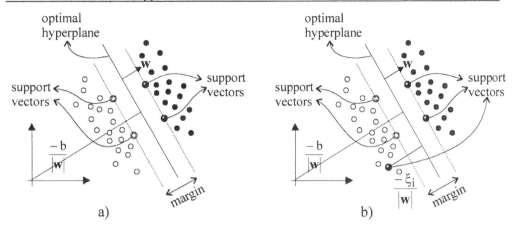

Figure 3.2. Optimal separating Hyper-plane (two-dimensional case (a) with separable data and (b) with non-separable data)[adapted from Burges, 1998].

One can find many possible linear classifiers that can separate the data, but there is only one that maximises the 'margin' (i.e. maximise the distance between it and the nearest data point of each class). This linear classifier is termed the *optimal separating hyperplane* (see Fig. 3.2.). It has also been shown that the optimal hyperplane, defined as the one with the maximal margin of separation between the two classes, has the lowest *capacity* (Vapnik, 1979).

The hyperplane $(\mathbf{w} . \mathbf{x}) + b = 0$ satisfies the conditions:

$$(\mathbf{w.x}_i) + b > 0 \quad if \ y_i = 1 \qquad and \qquad (\mathbf{w.x}_i) + b < 0 \quad if \ y_i = -1 \tag{3.5}$$

By combining the two expressions in Equation (3.5) and scaling $\mathbf{w}$ and b with an appropriate factor, an equivalent decision surface can be formulated as one that satisfies the constraint:

$$y_i[(\mathbf{w.x}_i) + b] \geq 1, \quad i = 1, \dots, l \tag{3.6}$$

and the hyperplane that optimally separates the data into two classes can be shown to be the one that minimises the functional:

$$\Phi(\mathbf{w}) = \frac{\|\mathbf{w}\|^2}{2} \tag{3.7}$$

Therefore, the optimisation problem can be reformulated as an equivalent non-constrained optimisation problem using Lagrangian multipliers, and its solution is given by identifying the saddle point of the Lagrange functional (Minoux, 1986), as follows:

$$L(\mathbf{w}, b, \alpha) = \|\mathbf{w}\|^2 / 2 - \sum_{i=1}^{l} \alpha_i \{[(\mathbf{w} \cdot \mathbf{x}_i) + b] y_i - 1\} \tag{3.8}$$

where the α_i are the Lagrange multipliers. The Lagrangian has to be minimised with respect to $\mathbf{w}$ and b, i.e:

$$\partial L / \partial b = 0 \quad \Rightarrow \quad \sum_{i=1}^{l} \alpha_i y_i = 0 \quad \textbf{and} \quad \partial L / \partial \mathbf{w} = 0 \quad \Rightarrow \quad \mathbf{w}_0 = \sum_{i=1}^{l} y_i \alpha_i \mathbf{x}_i$$

Substituting the above expression into Equation (3.8) will result in to the following dual form of the function which has to be maximised with respect to the constraints $\alpha_i \geq 0$:

$$W(\alpha) = \sum_{i=1}^{l} \alpha_i - (1/2) \sum_{i=1}^{l} \sum_{j=1}^{l} \alpha_i \alpha_j y_i y_j (\mathbf{x}_i \cdot \mathbf{x}_j) \tag{3.9}$$

The SVM offers an efficient way to improve computational and generalisation performance in a high-dimensional input space owing to this dual representation of the machine in which the training patterns always appear in the form of dot products between pairs of examples. Finding the solution of Equation (3.9) for real-world problems will usually require the application of quadratic programming (QP) optimisation techniques and numerical methods. Once the solution has been found in the form of a vector $\alpha^0 - (\alpha^0_1, \alpha^0_2, \dots, \alpha^0_l)$, the optimal separating hyperplane is given by

$$\mathbf{w}_0 = \sum_{supportvectors} y_i \alpha_i^0 \mathbf{x}_i \quad \text{and} \quad b_0 = -(1/2)\mathbf{w}_0 \cdot [\mathbf{x}_r + \mathbf{x}_s]$$

where $\mathbf{x}_r$ and $\mathbf{x}_s$ are any support vectors from each class. The classifier can then be constructed as:

$$f(\mathbf{x}) = sign(\mathbf{w}_0 \cdot \mathbf{x} + b_0) = sign\left(\sum_{support\ vector} y_i \alpha_i^0 (\mathbf{x}_i \cdot \mathbf{x}) + b_0 \right) \tag{3.10}$$

Only the points x_i which have non-zero Lagrangian multipliers α_i^0 are termed *Support Vectors* (SVs). If the data is linearly separable, all the SVs will lie on the margin and hence the number of SVs can be very small.

The above solution only holds for separable data, and still has to be slightly modified for non-separable data by introducing a new set of variables $\{\xi_i\}$ that measure the amount by which the constraints are violated (see Fig. 3.2b). Then the margin can be maximised by paying a penalty proportional to the amount of constraint violation. Formally, then one solves the following problem:

Minimise $\qquad \Phi(\mathbf{w}) = \|\mathbf{w}\|^2/2 + C\left(\sum \xi_i\right)$ $\qquad\qquad\qquad\qquad$ (3.11)

subject to $\qquad y_i[(\mathbf{w.x}_i) + b] \geq 1 - \xi_i, \quad and \quad \xi_i \geq 0 \qquad i = 1,...,l$

where C is a parameter chosen a priori and defining the cost of constraint violation. The first term in Equation (3.11) provides a minimisation of the VC-dimension of the learning machine, thereby minimising the second term in the bound of Equation (3.3). On the other hand, minimising the second term in Equation (3.11) controls the empirical risk, which is the first term in the bound of Equation (3.3). This approach, therefore, constitutes a practical implementation of Structural Risk Minimisation on the given set of functions. In order to solve this problem, the Lagrangian is constructed as follows:

$$\mathbf{L}(\mathbf{w}, b, \alpha) = \|\mathbf{w}\|^2/2 + C\left(\sum_{i=1}^{l}\xi_i\right) - \sum_{i=1}^{l}\alpha_i\{[(\mathbf{w}\cdot\mathbf{x}_i) + b]y_i - 1 + \xi_i\} - \sum_{i=1}^{l}\gamma_i\xi_i \qquad (3.12)$$

where α_i and γ_i are associated with the constraints in Equation (3.11) and the values of α_i have to be bounded as $0 \leq \alpha_i \leq C$. Once again, the solution of this problem is determined by the saddle point of this Lagrangian in a way similar to situation encountered in the case of separable data.

In the case where a linear boundary is inappropriate (or when the decision surface is non-linear), the SVM can map the input vector $\mathbf{x}$ into a higher dimensional feature space $\mathbf{z}$, by choosing a non-linear mapping a priori. Then the SVM constructs an optimal separating hyperplane in this higher-dimensional space. In this case, the optimisation problem of Equation (3.9) becomes:

$$W(u) = \sum_{i=1}^{l} u_i - (1/2)\sum_{i=1}^{l}\sum_{j=1}^{l} u_i u_j y_i y_j \mathbf{K}(\mathbf{x}_i \cdot \mathbf{x}_j) \qquad (3.13)$$

where $\mathbf{K}(x,y)$ is the kernel function performing the non linear mapping into the feature space, and the constraints are unchanged. Solving the above equation determines the Lagrange multipliers, and a classifier implementing the optimal separating hyperplane in the feature space is given by

$$f(\mathbf{x}) = sign\left(\sum_{i=1}^{l} y_i \alpha^0 \mathbf{K}(\mathbf{x}_i . \mathbf{x}) + b_0\right) \qquad (3.14)$$

Consequently, everything that has been derived concerning the linear case is also applicable for a non-linear case by using a suitable kernel $\mathbf{K}$ in addition to the dot product. Moreover, by using different kernel functions, the SV algorithm can be used to construct a variety of learning machines (see Fig. 3.3.), some of which appear to be similar to other more classical learning systems. Polynomial, radial basis functions, linear splines and certain sigmoid functions are among the candidates for acceptable kernels, and the corresponding mappings are described as follows:

- The simple polynomial kernel $\qquad \mathbf{K}(\mathbf{x}, \mathbf{x}_i) = ((\mathbf{x} . \mathbf{x}_i) + 1)^d$ where the degree of the polynomial d is user-defined

- Radial Basis Function kernel $\quad \mathbf{K}(\mathbf{x}, \mathbf{x}_i) = \exp(-\gamma|\mathbf{x} - \mathbf{x}_i|^2)$ where γ is user-defined

- Linear splines kernel: $\qquad \mathbf{K}(\mathbf{x},\mathbf{x}_i) = 1 + \mathbf{x}.\mathbf{x}_i + \sum_{j=1}^{n}(x_j - t_j)(x_{ij} - t_j)$ where
$(x_j - t_j) = 0$ if $x_j \le t_i$ and $(x_{ij} - t_j) = 0$ if $x_{ij} \le t_j$

- Neural network kernel : $\qquad \mathbf{K}(\mathbf{x}, \mathbf{x}_i) = \tanh(b(\mathbf{x} . \mathbf{x}_i) - c)$ where b and c are user-defined

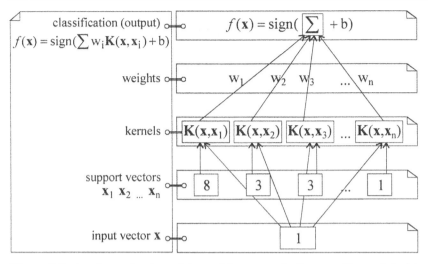

Figure 3.3. A schematisation of the architecture of a Support Vector Machine.

From the above one can say that, artificial neural network classifiers can in fact be derived as specific cases of SVMs corresponding to specific types of kernel functions.

The other case arises when the data lie in multiple classes. In order to obtain a k-class classification, a number of binary classifiers f^1, ..., f^k are constructed, each trained to separate one class from the rest, and these are combined by carrying out a multi-class classification (such as by applying a voting scheme) according to the maximal output before applying the sign function (Scholkopf, 1997).

3.3   Support Vector Regression

The Support Vector method can also be applied to regression problems, maintaining all the main features that characterise the maximal margin algorithm. As in the case of classification, SV regression tries to optimise the generalisation bound by the introduction of an alternative loss function that ignore errors that are within a certain distance of the true value (Cristianini and Shawe-Taylor, 2000). This type of function is reffered to as an ε-insensitive loss function.

Let the observed variable y take a real value, and let $f(\mathbf{x}, \alpha)$, $\alpha \in \Lambda$, be a set of real functions that contains the *regression function* $f(\mathbf{x}, \alpha_0)$. Considering the problem of approximating the set of data, $\{(\mathbf{x}_1, y_1), \ldots (\mathbf{x}_l, y_l), \quad \mathbf{x} \in \mathbf{R}^N, \quad y \in \mathbf{R}\}$ with a linear function, $f(\mathbf{x}, \alpha) = (\mathbf{w} \cdot \mathbf{x}) + b$, the optimal regression function is given by minimising the empirical risk:

$$R_{emp}(\mathbf{w},b) = \frac{1}{l}\sum_{i=1}^{l}|y_i - f(\mathbf{x}_i,\alpha)|_\varepsilon \tag{3.15}$$

With the most general loss function with an ε-insensitive zone described as

$$|y - f(\mathbf{x},\alpha)|_\varepsilon = \{\varepsilon \quad \text{if} \quad |y - f(\mathbf{x},\alpha)| \leq \varepsilon; \quad |y - f(\mathbf{x},\alpha)| \quad \text{otherwise} \} \tag{3.16}$$

the objective becomes one of finding a function $f(x, \alpha)$ that has at most a deviation of ε from the actual observed targets y_i for all the training data, and at the same time is as flat as possible. This is equivalent to minimising the functional

$$\Phi(\mathbf{w},\xi^*,\xi) = \|\mathbf{w}\|^2/2 + C\left(\sum \xi_i^* + \sum \xi_i\right) \tag{3.17}$$

where C is a pre-specified value which determines the trade-off between the flatness of f (x, α) and the amount up to which the deviation can be tolerated. ξ^*, ξ are slack variables representing upper and lower constraints on the outputs of the system (see Fig. 3.4.), provided as follows:

$$y_i - ((\mathbf{w}.\mathbf{x}_i) + b) \leq \varepsilon + \xi_i \qquad i = 1,2,...,l$$
$$((\mathbf{w}.\mathbf{x}_i) + b) - y_i \leq \varepsilon + \xi_i^* \qquad i = 1,2,...,l \tag{3.18}$$
$$\xi_i * \geq 0 \qquad \text{and} \qquad \xi_i \geq 0$$

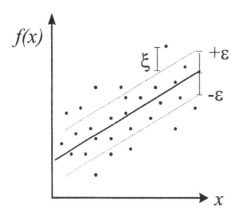

Figure 3.4. Pre-specified accuracy ε and a slack variable ξ in SV regression.

Now the Lagrange function is constructed from both the objective function and the corresponding constraints by introducing a dual set of variables, as follows:

$$L = \|\mathbf{w}\|^2/2 + C\left(\sum_{i=1}^{l}(\xi_i + \xi_i^*)\right) - \sum_{i=1}^{l}\alpha_i[\varepsilon + \xi_i - y_i + (\mathbf{w}.\mathbf{x}_i) + b]$$
$$- \sum_{i=1}^{l}\alpha_i^*[\varepsilon + \xi_i^* + y_i - (\mathbf{w}.\mathbf{x}_i) - b] - \sum_{i=1}^{l}(\eta_i\xi_i + \eta_i^*\xi_i^*)$$

(3.19)

It follows from the saddle point condition that the partial derivatives of L with respect to the primary variables ($\mathbf{w}$, b, ξ_i, ξ_i^*) have to vanish for optimality. Substituting the results of this derivation into Equation (3.17) yields the dual optimisation problem:

$$W(\alpha^*, \alpha) = -\varepsilon\sum_{i=1}^{l}(\alpha_i^* + \alpha_i) + \sum_{i=1}^{l}y_i(\alpha_i^* - \alpha_i) - \frac{1}{2}\sum_{i=1}^{l}\sum_{j=1}^{l}(\alpha_i^* - \alpha_i)(\alpha_j^* - \alpha_j)(\mathbf{x}_i.\mathbf{x}_j)$$

(3.20)

that has to be maximised subject to the constraints:

$$\sum \alpha_i^* = \sum \alpha_i \; ; \qquad 0 \leq \alpha_i^* \leq C \quad \text{and} \quad 0 \leq \alpha_i \leq C \qquad \text{for } i = 1, 2, \ldots, l\,.$$

The solution of Equation (3.20) provides the coefficients α_i^* and α_i for all $i = 1$ to l and it can be shown that all the training patterns within the ε-insensitive zone yields α_i^* and α_i as zero (Vapnik, 1998). The remaining training patterns constitute the Support Vectors and the approximating function is given by:

$$f(\mathbf{x}) = \sum_{supportvectors}(\alpha_i^* - \alpha_i)(\mathbf{x}_i.\mathbf{x}) + b_0$$

(3.21)

where $b_0 = -(1/2)\mathbf{w}_0.[\mathbf{x}_r + \mathbf{x}_s]$ and $\mathbf{w}_0 = \sum_{supportvectors}(\alpha_i^* - \alpha_i)\mathbf{x}_i$

Once again, when linear regression is not appropriate, as in the case of most engineering applications, a non-linear mapping kernel $\mathbf{K}$ is used to map the data into a higher-dimensional feature space where linear regression can be performed. Once the optimum values α_i^0 and α_i^{0*} are obtained, the regression function is given by:

$$f(\mathbf{x}) = \sum_{support\ vectors}(\alpha_i^{0*} - \alpha_i^0)\mathbf{K}(\mathbf{x}_i, \mathbf{x}) + b_0$$

(3.22)

where $b_0 = -(1/2) \sum_{sup\ port\ vectors} (\alpha_i^{0*} - \alpha_i^0)[\mathbf{K}(\mathbf{x}_r, \mathbf{x}_i) + \mathbf{K}(\mathbf{x}_s, \mathbf{x}_i)]$

The possibility of using Support Vector Machine as a generalisation of most model induction methods will be demonstrated in the following section by applying the technique on two case studies of hydrologic and hydrodynamic modelling problems.

3.4   Applications of Support Vector Machines

Although the use of Support Vector methods in application has begun only quite recently, a considerable number of researchers have already reported state of the art performances in a variety of applications, such as to pattern recognition, regression estimation, and time series prediction (Muller et al., 1997). In most cases, SVMs have been shown to deliver a very high performance in a number of real world applications (Cristianini and Shawe-Taylor, 2000). Babovic et al. (2000) applied SVMs for water level forecasting in the city of Venice and obtained consistently better results than those obtained using ANNs. Dibike (2000), Sivapragasam and Liong (2000) and Sivapragasam et al. (2001) have also demonstrated that the SVM provides good performance in terms of generalisation for rainfall-runoff modelling problems. The following two case studies presented here also further demonstrate the applicability of SVMs for higher dimensional regression estimation (modelling) problems in hydrology and in hydraulic engineering and compare the performance with that of ANNs and other conceptual or empirical models.

3.4.1   SVMs for rainfall-runoff modelling

Determining the relationship between rainfall and runoff for a catchment is one of the most important problems faced by hydrologists and engineers (ASCE, 2000b). Information about stream flow is usually needed for engineering design and management purposes. Although some catchments have been gauged to provide continuous record of stream flow, engineers are often faced with situations where little or no stream flow information is available. Hence, the availability of extended records of rainfall and other climatic data, which can be used to estimate stream flow data, is the major reason for the origin of rainfall-runoff modelling (Lorrai, 1995). This relationship between rainfall and runoff is known to be highly non-linear and complex. In addition to the mean rainfall over an area, runoff is dependent on numerous factors such as initial soil moisture, land use, catchment geomorphology, evaporation, distribution and duration of rainfall over the area, and so on. In this context, models, which provide a physically sound description of hydrological processes occurring in a catchment, may have some advantages over purely empirical ones. However, physically based models requires a more detailed understanding of all the processes involved and they need a large amount of site-specific data for their proper implementation. Therefore, in operational hydrology two alternative approaches are usually considered for rainfall-runoff modelling. One is the conceptual modelling method and the other is the systems analysis or data-driven modelling approach.

Conceptual rainfall-runoff models are designed to approximate within their structures the general internal sub-process and physical mechanisms that govern the hydrologic cycle. While conceptual models may be of importance in the understanding of hydrologic processes, there are many practical situations of stream flow simulation and forecasting where the main concern is that of making accurate predictions of flows at specific locations corresponding to different rainfall scenarios that are anticipated to occur in the catchment. In such situations, a hydrologist may prefer to implement a simpler data-driven model to identify a direct mapping between the inputs and outputs without detailed consideration of the internal structure of the physical processes (Hall and Minns, 1993).

Artificial Neural Networks (ANNs) are by now the most popular non-linear data-driven modelling techniques that have been successfully applied as means of estimating the runoff response of a catchment on the basis of known meteorological time series of such processes as rainfall and evaporation (see Smith and Eli, 1995; Minns and Hall, 1996; Dibike, 2000). In this study, however, the possibility of using Support Vector Machines (SVMs), for rainfall-runoff modelling is investigated by applying the method to three catchments of different size and different rainfall intensity (see Table 3.1.). The performances of the SVM is then compared with that of the ANN and previously published results of a modelling of the same catchments with a conceptual model referred to as SMAR (see Kachroo et al., 1995).

SMAR is a conceptual Soil Moisture Accounting and Routing rainfall-runoff model developed by the Department of Hydrology, University College, Galway, Ireland. The water balance component in this model operates in a manner analogous to a vertical stack of horizontal soil layers each of which can contain a certain amount of water at field capacity (Kachroo, 1992). Evaporation occurs from the top layer at the potential rate, and from the second layer, only on the exhaustion of the first, at the remaining potential multiplied by a parameter a, whose value is less than unity. On exhaustion of the second layer, evaporation from the third layer occurs at the remaining potential multiplied by a^2 and so on. Thus, a constant potential evaporation applied to the basin would reduce the soil moisture storage in a roughly exponential manner. This model has five parameters that should be optimised during calibration under the constraint that they are optimised at a reasonable value.

When applying data-driven models to the rainfall-runoff transformation, the stimulus is obviously the rainfall, and the response is the stream flow at the basin outlet. Since the flow at any instant is effectively composed of contributions from different sub-areas whose time of travel to the outlet covers a range of values, both the concurrent and antecedent rainfalls should be considered as stimuli to the data-driven models.

Catchment Description

The data sets used in this study were taken from a report on a flood forecasting workshop conducted in 1995 in Galway, Ireland, where intercomparison studies were made on mathematical models for river flow forecasting (Kachroo et al., 1995). Three catchments of different size and climatic conditions were identified (see Table 3.1.). The first catchment, Baihe, is located in China and has a relatively large area. This catchment is semidry with a mean rainfall intensity of 2.6mm/day and a mean discharge intensity of 1mm/day over the catchment. The second catchment, Chu, is located in Vietnam and has a medium size and a relatively wet climate with mean rainfall intensity of 3.9mm/day producing a mean discharge intensity of 1.7mm/day. The third catchment, Balephi, is located in Nepal and covers a relatively small area. It has a very wet climate with a mean rainfall intensity of 9.3mm/day and mean discharge intensity of 7.2mm/day. The average stream flow/rainfall ratio of the three catchments is 0.38, 0.44 and 0.77 respectively. Eight years of data were available for each catchment and the first 5-year records were used for calibration while the remaining 3-year records were employed for validation.

Table 3.1. Climatic characteristics of catchments considered in the second case study.

Catchment name	Country	Catchment area (Km²)	Mean rainfall (mm/day)	Mean pan evap. (mm/day)	Mean disch. (mm/day)
Baihe	China	61,780	2.6	2.8	1.0
Chu	Vietnam	2,090	3.9	2.6	1.7
Balephi	Nepal	630	9.3	3.6	7.2

Data Preparation

The daily rainfall, evaporation and stream-flow data from each of the three catchments was pre processed while taking some physical understanding of the rainfall-runoff process into account. Previous experience has shown that catchment responses are usually influenced by antecedent conditions that can extend over several days, weeks, or even months, back into the past. Therefore, in addition to the current and a few previous days of rainfall and evaporation data, the average of the current and previous '*n*' days (where '*n*' typically took values such as 5, 10, ...) of rainfall and evaporation data were prepared as input to each of the data models, while the current discharge was taken as the output from these models. A large number of experiments were conducted to identify the appropriate lag times for rainfall and evaporation data. This finally resulted in a data set with 17 inputs (The present and each of the last 5 days of rainfalls; the averages of the last 5, 10, 20, 30, 50 and 100 days of rainfall; the present pan evaporation and the averages of the last 5, 15, 30 and 100 days of pan evaporation) and one output (the present-time runoff). In this arrangement, while the '*n*' days average rainfall provided information about the state of the basin, which represented the capability of the river system to respond to rainfall perturbations, the daily rainfall values gave a measure of the amount of water recently gathered by the basin and represented the perturbation experienced by the river system. No antecedent values of runoff were included as input to the ANNs and SVMs, resulting in simulation models rather than forecast routines.

One other important factor in machine learning is the setting up of the appropriate parameters for the learning machines. When applying SVMs, in addition to the specific kernel parameters, the optimum values of the capacity factor C and the size of the error-insensitive zone ε should be determined during the modelling experiment. For ANNs, in addition to the different learning parameters, the optimal number of nodes in the hidden layer should also be determined in order to reach a less complex network with a relatively good simulation capability.

Results and Discussion

The best simulation performances of SVM and ANN, along with the previously reported performance of the conceptual model SMAR, are summarised in Table 3.2. The table contains the RMS errors observed during the training and verification periods, corresponding to each model type as applied to each one of the three catchments.

Table 3.2. Performances of the different types of rainfall-runoff models over the training and verification data.

Catchment	Training RMSE			Verification RMSE		
	ANN	SVM	SMAR	ANN	SVM	SMAR
Baihe	0.406	0.433	0.812	0.869	0.619	0.916
Chu	1.845	2.434	2.744	2.902	2.162	2.762
Balephi	1.701	2.311	3.351	4.459	3.920	4.543

In the case of SVM training, three types of kernel functions, namely, Polynomial Kernel, Radial Basis Function Kernel and Neural Network Kernel were used. Moreover, the parameter ε corresponding to the error insensitive zone, the capacity factor C and other kernel-specific parameters had to be set to their optimal values during the model identification (training) process. At the moment, identification of the optimal values for these parameters remains largely a trial and error process, which does however become much easier with some practice. Table 3.3. shows the minimum training and verification errors obtained with SVMs corresponding to the (near) optimal set of parameters for each of the above kernel functions. In most of the cases, a better performance was achieved with the radial basis function kernel.

Table 3.3. The different Kernel functions used during SVM training and the corresponding performances.

	Polynomial Kernel $K(x,y) = ((x \cdot y) + 1)^d$		Radial Basis Kernel $K(x,y) = \exp(-\gamma|x - y|^2)$		Neural Network Kernel $K(x,y) = \tanh(b(x \cdot y) - c)$	
RMSE/ Catchments	Training	Testing	Training	Testing	Training	Testing
Baihe	0.549	0.811	0.433	0.619	0.657	0.669
Chu	2.434	2.162	1.933	2.358	2.451	2.583
Balephi	2.527	4.298	2.311	3.920	2.665	4.419

The values in Table 3.2. show that there was no major difference in performance (expressed here in terms of the RMSE values) between the different modelling approaches investigated in this study. However, in general, the data-driven models exhibited slightly better performances than the conceptual ones. Moreover, as compared to the ANN, the SVM seemed to achieve on average a 15% increase of accuracy in estimating the runoff for the verification period. In Figures 3.5 (a, b, c), the SVM simulated runoff is plotted against the measured values for one out of the three years of the verification period (for clarity of presentation) for each of the three catchments. Once again this figure shows an acceptable fit between the measured and simulated plots of runoff values for each of the catchments considered. The models seem to have captured the main features of the hydrographs and even the peak flows are simulated reasonably well.

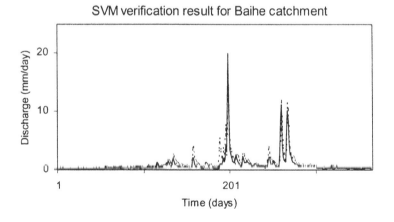

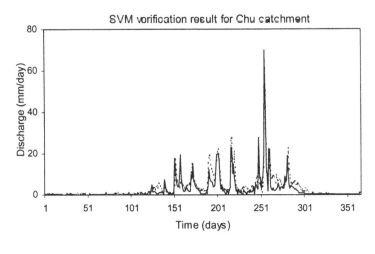

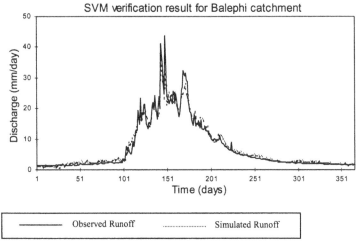

Figure 3.5 (a, b, c). Rainfall-runoff Simulation results with SVM applied on test data plotted against the measured values.

3.4.2   SVMs for the prediction of a horizontal force on a vertical break water

Vertical breakwaters are predominantly upright concrete structures built in the sea in order to protect an area from wave attack. They are specifically used in harbour areas with relatively high water depths, where traditional rubble-mound (or rock slope) breakwaters become a rather expensive alternative. The stability analysis of such vertical breakwaters (as in Fig. 3.6.) relies on an estimation of the total force and overturning moments caused by dynamic wave- pressure action (van der Meer and Franco, 1995). Therefore, the prediction of the total horizontal force exerted on the breakwater by the waves is one of the necessary steps in the stability analysis of the structure.

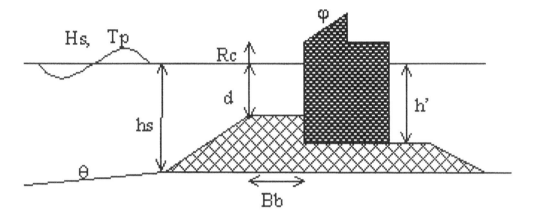

Figure 3.6. Vertical breakwater layout and the different parameters involved [adapted from Gent and Boogaard, 1998].

Present-day design practice for vertical breakwaters mainly depends on physical model tests and empirical formulae. However, the physical processes involved are so complex that their representation in imperial relationships is not always very clear. One of the best known sets of empirical formulae to predict total horizontal forces on vertical structures is provided by the Goda method (Goda 1985). Goda's formula calculates the distribution of wave pressures on a vertical structure based on a knowledge of structure geometry, seabed characteristics and wave parameters in front of the structures. However, subsequent studies have shown that the horizontal forces calculated by Goda's formula are up to 10% higher than the corresponding measured force (van der Meer and Franco, 1995). Gent and Boogaard (1998) has also reported that for the data set used in this study, the higher forces are under-predicted by the Goda-method and the reliability of this formula has by far the largest influence on the calculation of the probability of failure of the breakwater structures.

The total horizontal force on a vertical breakwater is the result of the interaction between the wave field, the foreshore and the structure itself. The wave field in the offshore region is usually described using only two parameters, namely, the significant wave height in the offshore region (H_s) and the corresponding average period between peak waves (T_p). The effect of the foreshore is described by the slope of the foreshore (tan θ) and the water depth in front of the structure (h_s). The structure is characterised by the height of the vertical wall below and above the water level (h' and R_c), the water depth above the rubble-mound foundation (d) and the shape of the superstructure (as characterised by the angle φ). The berm-width is also added as an additional parameter (B_b).

The data used for this study were collected from physical-model tests performed at several hydraulic laboratories in Europe. During the experiments, wave trains with lengths of 1000-3000 waves were considered and, based on the published data, a parameter corresponding to the total horizontal force with exceedance frequency of 0.4% were derived (Meer and Franco, 1995; Gent and Boogaard, 1998). In total, 612 data-patterns obtained from various hydraulic research institutes in different European countries were collected. In previous studies, Gent and Boogaard (1998) used this data set to train artificial neural networks (ANNs) for the successful prediction of the horizontal force on vertical breakwaters. They also followed a slightly modified approach, by including explicit physical knowledge adopted from Froude's law, in preparing the data for training the network. The input/output patterns were thereby re-scaled to a form where the H_{s0} (equivalent deep-water wave height) become 0.1m (this choice of 0.1 being rather arbitrary). As a result, every input output pattern was scaled with a different factor λ. In this way a 'new' data set, still consisting of 612 input/output patterns, was produced but the input dimension was reduced from 9 to 8 by removing H_{s0}. Gent and Boogaard reported that, as a result of this reduction of the input dimension, in addition to the important physical knowledge included in the data, training could be performed with an ANN of reduced size and complexity.

In this study, however, the possibility of using Support Vector Regression (SVR) as yet another data-driven modelling approach for the prediction of the horizontal force on a vertical breakwater has been investigated. The same data set described above is used in order to make comparison with the previous ANN modelling results. A data set of 500 input-output patterns was randomly selected to train the SVMs and the remaining 112 were used for verification. Four types of kernel functions, namely the simple dot product, simple polynomial, radial basis function and linear splines were used for the SVM training. Using the simple dot product kernel amounts to approximating with a linear

SVM, while the rest of the kernel results in non-linear SVMs. Table 3.4. shows the RMSE of best performing SVMs for each kernel type, with the corresponding numbers of support vectors. The relatively poor performance of the dot product kernel is an indication of the non-linearity of the relationship between the dependent and independent variables implying that non-linear kernel function have to be used to approximate the relation. When radial basis function and linear splines kernels were used to train the SVMs, the verification results were found to be comparable with those of the ANN. Since the data set on which both the SVMs and ANNs trained contained some inconsistencies, the very small differences in the verification results do not justify the statement that one method is better than the other. At the moment, identifying the optimal values of C, ε and other kernel specific parameters is largely a trial and error process (see also Sivapragasam et al., 2001). However, an increase in the capacity of the machine usually provides an increase in the number of corresponding support vectors. In addition to the risk of overfitting, this may result in a relatively longer computational time and increased memory requirement. There should, therefore, always be a trade off between these factors, and this also holds in the case of ANN applications. In the present application, the SVMs were found to generalise well by setting the capacity factor C between 10 and 100. Similarly, in most cases, ε values in the range of 0.01 to 0.025 resulted in the best performing SVM. For radial basis function kernels, γ values in the range of 3 to 10 were found to be appropriate. So far, there does not appear to be any rational basis for these parameters.

Table 3.4. General performance of SVMs with different kernel functions.

Kernel function	Training Performance		Testing Performance	
	RMSE	MAE	RMSE	MAE
simple dot product	0.311	0.207	0.288	0.195
simple polynomial	0.218	0.139	0.219	0.152
radial basis function	0.208	0.111	0.199	0.121
linear splines	0.211	0.119	0.191	0.126
ANN (with sigmoid tran. fun.)	0.203	0.133	0.186	0.129

As can be seen from Table 3.4., the overall performance of the SVM (especially with radial basis kernels) is quite comparable with that of ANN's, with mean absolute errors (both in training and testing) of less than 10% of the data range. However one can also see that the SVM predicted the output quite well in the lower and middle range of the data set while the ANN seemed to predict the higher range of the output values relatively better (see Fig. 3.7.). One can observe from the test results in Table 3.4. that while the root mean square errors (RMSEs) corresponding to SVMs with radial basis and linear-spline kernels are comparable with those of ANNs, they also provide a slightly lesser value of MAE. However, in both cases, a large scatter is observed, which is to a large extent caused by the quality of the measured data (see Gent and Boogaard, 1998). The overall results of this investigation show that SVMs are still able to provide realistic predictions of forces on vertical structures. moreover, the performances of the SVMs shown in Table 3.4. could still be improved by exploring even more combinations of values for the capacity and kernel specific parameters during the training stage.

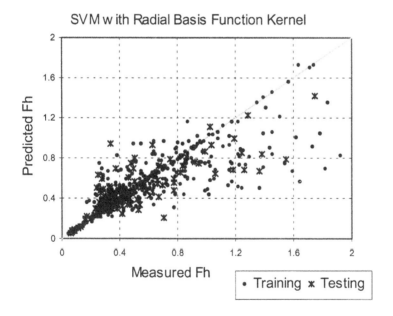

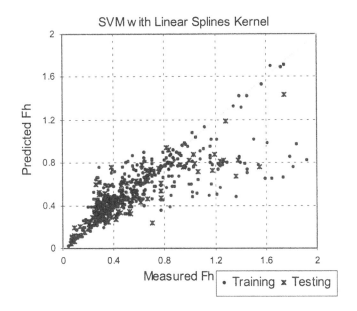

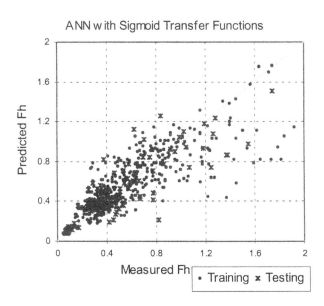

Figure 3.7(a, b, c). SVMs and ANN performances in the prediction of horizontal force on a vertical break water.

3.5   Discussion

This study has, in general, shown the applicability of the principles of Support Vector Machines (SVMs) and their underlying statistical learning theory as a new approach for model induction from data. Since the SVM is largely characterised by the type of its kernel function, it is necessary to choose the appropriate kernel for each particular application or problem in order to guarantee satisfactory results. It has also been emphasised that, using some physical insight into the real world problem while preparing the input and output data sets could help in simplifying the model identification process and in finding the best generalising model.

The SVM approach for rainfall-runoff modelling has been demonstrated to provide a good alternative to the traditional use of conceptual modes and has even performed much better than the ANN. The modelling experiment has shown that the structural risk minimisation induction principle gives the SVM the important property of generalising an performing well in the presence of unseen data. The SVM has also demonstrated a reasonably satisfactory performance for the prediction of horizontal forces on vertical structures due to dynamic waves. In this particular case, the performance of the SVM was comparable to that of ANNs trained with sigmoid transfer functions. The lack of a significant improvement in the generalising ability of the SVM over the ANN could be attributed (in this particular case) to the presence of some inconsistenies in the data set collected from different hydraulic laboratories throughout Europe (Gent and Boogaard, 1998). To cope with this problem, Boogaard et al. (2000a) has proposed a method of repeated experiment with resembling to get more insight in to the uncertainty of such models so that for a given input data, one can provide the corresponding output along with the confidence interval associated with it.

In general, SVMs provide an attractive approach to data modelling and this study shows its potential as an alternative modelling technique for applications in Hydraulics and Hydrology. Especially when modelling input-output patterns with a high-dimensional input space, SVMs offer an efficient way to deal with the computational and generalisation problems due to the fact that the training patterns always appear in the form of dot products between pairs of examples and the optimisation problem has a unique solution. However, despite their encouraging performance in this and other similar studies, several aspects of SVMs still remain to be addressed. For example, determining the proper parameters for the capacity factor C and the size of the error insensitive zone ε is still a heuristic process and the automation of this process would be beneficial. The other limitations are those associated with a relatively long computational time required to find the optimum solution and the relatively large increase in memory requirement with the increase in the size of the training set.

Chapter 4

Artificial Neural Networks as Domain knowledge Encapsulators

4.1   Introduction

Since the early nineties, artificial neural networks have been successfully used in hydrology-related areas such as rainfall-runoff modelling, stream flow forecasting, groundwater modelling, water quality studies, precipitation forecasting and reservoir operations (ASCE, 2000). One of the appealing features of ANNs is their ability to acquire and generalise *knowledge* by learning from data. However, despite the powerful processing capabilities of neural network systems, interpretations of their internal knowledge representation are largely inaccessible to humans. Neural networks do not provide an 'explanation' as part of their information processing; rather, the 'knowledge' that they have gained through training is stored in their weights. Until recently, it was a widely accepted idea that neural networks were 'black boxes', that is, the knowledge stored in their weights after training was not accessible to any meaningful inspection, analysis, and verification (Omlin and Giles, 2000). But, nowadays, explicit modelling of the knowledge represented by neural network systems is becoming an important area of research in *Knowledge-Based Neurocomputing* (Cloete I., 2000). The key assumption of knowledge-based neurocomputing is that knowledge is obtainable from, or can be represented by, a neural network system in a humanly-comprehensible form. A useful starting point for understanding this concept is that within a trained artificial neural network, *knowledge* acquired during the training phase is encoded as: (a) the network architecture, (b) the activation functions associated with each hidden and output units of the network and (c) the connection weights of the network. The task of extracting explanations from the trained artificial neural network is therefore one of interpreting in a comprehensible form the collective effect of (a), (b) and (c) (Andrews et al., 1995). This means that, the knowledge imbedded in the neurocomputing system is also representable in a more conventional symbolic or other familiar well-structured form, such as in the form of rules or partial differential equations.

Differential equations are very important mathematical tools in engineering. They provide the dominant techniques for modelling dynamic systems, because the causal relations between physical variables present in most engineering problems can often be described reasonably well by differential equations. However, if the dynamics of the physical system are unknown, then it will be difficult to identify the process (or differential equations) driving the system. This modelling problem can then be approached from the side of neural networks, where the network corresponds to the model and the network weights are the free parameters. A learning algorithm searches the optimal network weights that fit the network output to the actual system observation. Once learning is accomplished and the final optimal weights are obtained, the trained neural network will be ready to simulate the behaviour of the real system. Therefore, understanding the knowledge imbedded in neural network systems (implicitly distributed in the weighs) is important because it gives more confidence to engineers in their application of these devices, while this confidence can make the technique more useful in many areas of application. Researches in the topic have resulted in a number of algorithms for extracting knowledge in symbolic form from trained neural networks (Minns, 1998; Omlin and Giles, 2000). The present work also tries to demonstrate this feature in the domain of hydraulics by showing that ANNs can encapsulate the same knowledge as do continuum equations by analysing the weights of the ANNs that are trained with synthetic data and actually restoring the continuum partial differential equations (PDEs).

4.2   Applications of Artificial Neural Networks to the Generation of Wave Equations from Hydraulic Data

For the purpose of encapsulating and processing knowledge, artificial neural networks appear among the most popular and most-extensively developed of data-driven modelling techniques (see Minns, 1998). However, it is once again only natural that there should remain a certain skepticism within the hydraulics community about the ability of artificial neural networks to encapsulate knowledge that has been 'traditionally' encapsulated in the form of continuum (and, in this case, partial-differential) equations. A number of researches (Babovic and Abbott, 1997; Babovic and Keijzer 2000; Dibike, 2000; Whigham and Crapper, 2001) demonstrated the possibility of evolving equations from hydraulic data using evolutionary algorithms like genetic programming. Genetic Programming (GP) can be used to evolve algebraic expressions using a given set of algebraic functional elements (Koza, 1992). It allows the optimisation of a symbolic expression using a tree structure representation. However, the first purpose of the present work is to demonstrate that ANNs can encapsulate the same knowledge, or exhibit the same semantic content, as do continuum equations. The argument is that, if numerical solutions of specific partial differential equations (that are already generally accepted as means for encapsulating our knowledge on the behaviour of specific aspects of nature) are used to provide the data sets that are to train the ANNs, and the resulting trained weights of the ANNs really are able to reinstate the original PDEs, then data taken from nature (that may be treated to minimise noise and other distorting influences) should just as well provide ANNs that can simulate the dynamics of the system under consideration.

The usual procedures of numerical modelling and those followed here, together with their relations in the present study, can be schematised in the form of a category as follows (see also Abbott and Dibike, 1998):

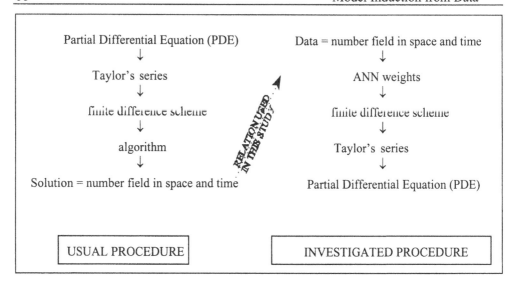

Most of the phenomena that are to be simulated in the next generation of computational engines, and which could make extensive use of the kind of approach investigated here (Abbott, 1997a), have to do with advection and related wave-like phenomena. A first trial application of this approach was made by Minns (1998) to the most simple of such processes, namely that of pure advection, as described by an advection, or 'scalar wave', equation. This equation describes the movement or 'transport' of any property of the fluid with a representative velocity or 'celerity', which may be a function of the fluid property itself. Minns (1998) showed that, in the simple case of pure advection with a constant velocity, a linear ANN is capable of learning the exact solution, which is also exactly equivalent to the differential equation description. Moreover, for problems of variable velocity, he also showed that non-linear ANNs (ANNs with sigmoid threshold function) are able to provide exceptionally accurate solutions over the range of velocities for which they were trained. Finally he indicated that even if the exact mathematical formulation of a physical process is known, the use of ANNs in the solution of these partial and ordinary differential equations may offer an improvement over the traditional methods of numerical analysis, such as finite difference methods, which are plagued by problems of instability and inaccuracy. In a similar vein, the present work investigates the application of artificial neural networks (ANNs) to one of the simplest of problems of short-period wave modelling.

4.2.1 Boussinesq equations for short-period waves

The transformation of short-period waves as they propagate in shallow water is a phenomenon of fundamental importance in hydraulic engineering. It is especially important for the design and analysis of engineering works in the coastal environment, and the corresponding wide range of applications has prompted the development of a variety of mathematical descriptions of the hydraulic phenomena involved. The most widely used, Boussinesq-like equations, are a modification of the well known long wave equations making some allowance for the effect of the non-hydrostatic pressure distribution that occurs in short-period wave propagation. They are correspondingly capable of describing the transformational behaviour of irregular, non-linear waves in relatively shallow water. It may be shown (e.g. Abbott et al., 1978) that the Boussinesq equations have a very wide range of application. The now 'classical' equations are based on shallow water assumptions and the practical deep-water limit of their applicability in the case of a horizontal bed is usually taken to correspond to a depth-to-deep-water wave-length ratio, h/L_0, equal to about 0.22. Newer equations (Dingemans, 1997), however, incorporate a number of Boussinesq correction terms and can be used up to a depth-to-deep-water wave-length ratio of 0.5 and even beyond These equations can provide good approximations to the behaviour of short-period waves in shallow waters even with ratios of wave height to wave length (wave steepness) up to the conditions of the breaking zone.

The simplest Boussinesq-like equations for a one-dimensional flow over a horizontal bed, while neglecting boundary shears, are given by the so-called *one-way* or *unidirectional* propagation models. One of the best known examples describing a balance between the non-linear and dispersive terms is that of the Benjamin, Bona and Mahony (BBM) equation (see Dingemans, 1997). This is usually given as:

$$\frac{\partial \zeta}{\partial t} + \frac{3}{2}\sqrt{\frac{g}{h_0}}\,\zeta\,\frac{\partial \zeta}{\partial x} + \sqrt{gh_0}\,\frac{\partial \zeta}{\partial x} - \frac{1}{6}h_0^2\,\frac{\partial^3 \zeta}{\partial x^2 \partial t} = 0 \qquad (4.1a)$$

where h_0 is the still-water depth and ζ is the elevation of the free surface with respect to the still-water level.

An earlier and still quite widely used form is that of the Kortewege de Vries (KdV) equation which reads:

$$\frac{\partial \zeta}{\partial t} + \frac{3}{2}\sqrt{\frac{g}{h_0}}\zeta \frac{\partial \zeta}{\partial x} + \sqrt{gh_0}\frac{\partial \zeta}{\partial x} + \frac{1}{6}h_0^2 \sqrt{gh_0}\frac{\partial^3 \zeta}{\partial x^3} = 0 \qquad (4.1b)$$

For the purpose of the present study, (4.1a) is further simplified by substituting the term $M\zeta$ by Mh (where h is the instantaneous depth of flow, and so assuming a horizontal bottom) to provide a third-order PDE of the general form:

$$\frac{\partial h}{\partial t} + C\frac{\partial h}{\partial x} + D\frac{\partial^2 h}{\partial x^2} - E\frac{\partial^3 h}{\partial x^2 \partial t} = 0 \qquad (4.2)$$

where C, D and E are coefficients corresponding to celerity, diffusion and frequency-dispersion terms. One may already anticipate from (4.1a) and (4.1b) that D should be zero; i.e. that all energy transport is due to dispersion.

4.1.2   Generation of Boussinesq-like short-period waves equations with ANNs

The search for even more accurate and still-stable solution techniques for solving the Boussinesq-like equations continues to be a subject of ongoing research in numerical modelling. This section, however, describes only an investigation into the application of artificial neural networks (ANNs) to the simplest of problems of short-period wave modelling. The 'standard' multi-layered perceptrons with back propagation appear as the simplest means available for this purpose. They can be interpreted alternatively as one- or two-level numerical schemes (of, in principle, arbitrary numerical accuracy) constructed between the uppermost and lowermost layers, from which the governing partial differential equations can be constructed. However, it must be emphasised that the main purpose of this study towards the derivation of numerical schemes and PDEs using ANNs is to induce some confidence in the predictive abilities of the ANNs so that their application can be widened to cover different areas of computational hydraulics.

The experiments in this investigation involved training an ANN to reproduce a solution of a Boussinesq-like equation on a finite grid in a way similar to that reported for the pure advection equation. The patterns used to train the network were generated using the

Boussinesq Wave (BW) module of the MIKE 21 modelling system of the Danish Hydraulic Institute. In this module, both the classical and the newer forms of the Boussinesq-like equations are available. The equations actually applied, when reduced to one-dimensional flow on a horizontal bed with no resistance, read as follows:

Continuity
$$\frac{\partial \zeta}{\partial t} + \frac{\partial p}{\partial x} = 0$$
(4.3a)

Momentum
$$\frac{\partial p}{\partial t} + \frac{\partial}{\partial x}(\frac{p^2}{h}) + gh\frac{\partial \zeta}{\partial x} + \psi = 0$$
(4.3 b)

where
$$\psi = -(B + \frac{1}{3})h^2 \frac{\partial^3 p}{\partial x^2 \partial t} - Bgh^3 \frac{\partial^3 \zeta}{\partial x^3}$$

B is a linear dispersion factor and p $(=uh)$ is the volume flux density in the x direction under the assumtion that the velocity u is uniform with depth.

It is shown by Madsen and others (see Madsen et al., 1991) that the classical Boussinesq equation corresponds to (4.3) with the value of $B=0$, while a newer form of Boussinesq equation that was incorporated in the MIKE 21 BW module took the value $B=1/15$. Although these equations are not entirely equivalent to (4.1a) or (4.1b), the differences were deemed negligible in the tests that were employed here.

4.3   Numerical Experiments

For the first set of numerical experiments, a rectangular channel 350m long and 5m wide in prototype was considered. In a first series of tests, only waves of constant form were introduced. The ANN then provided only a pure advection equation, which, although it reproduced the analytical celerity of the wave train, considerably underestimated the coefficients in the third derivative term over a wide range of tests. This result was not surprising, since the ANN had no opportunity to learn the redistribution effects of the third derivative term, and simply incorporated almost all of its influence in the celerity of the waves. It did, however, reproduce the analytical celerity to a high degree of accuracy over a considerable range of water depths. In view of this behaviour, all further tests, as reported here, were made with an irregular uni-directional short wave with a wave period of 6 seconds (L_0 . 56 m, where the deep-water wave length L_0 . $gT^2/2\pi$) having a significant wave height of 0.5 m, as shown in Figure 4.1. This was introduced as a boundary condition at one end, while, at the other end, the channel was closed and a sponge layer was set up to provide radiation boundary conditions which could absorb the wave energy propagating out of the model area.

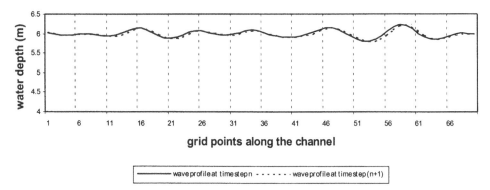

Figure 4.1. Wave profile at two consecutive time-steps in a channel with water depth of 6m.

The initial depth of water in the channel was first taken to be 6 m and the channel was divided into equal increments of $\Delta x = 5$ m, while a time interval of 0.5 second was adopted for the simulation. With this discretization, the maximum Courant number was approximately unity. For the given average depth of flow, the h/L_0 ratio corresponding to the wave with wave period of 6 sec was found to be 0.11; therefore, the classical form of the Boussinesq equation could be used during the simulation to generate the training data.

The input array to the ANN consisted of four values, representing the values of the water depth h at the grid points $j-2, j-1, j$, and $j+1$ at time level n, while the output consisted of the value of the water depth h at grid point j at time level $n+1$. The network was trained with wave profiles of five consecutive time steps (wave profiles of two consecutive time steps are shown in Fig. 4.1.) and then verified with another five. To begin with, a neural network with no hidden layer and with a linear activation function was considered. During training, the back-propagation algorithm converged to a residual RMS error of less than 0.001. The corresponding distribution of weights is shown in Figure 4.2.

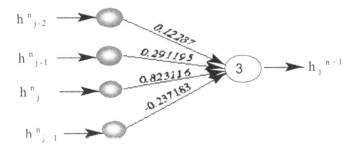

Figure 4.2. Distribution of weights in the simple neural network.

Since the activation function used in the above example is linear, the mathematical expression for the solution can be given as a simple weighted sum of the input variables, i.e.:

$$h_j^{n+1} = 0.12287\,h_{j-2}^n + 0.291195\,h_{j-1}^n + 0.823116\,h_j^n - 0.237183\,h_{j+1}^n \qquad (4.4)$$

The Taylor series expansion of the terms in (4.4) about the centre point of the scheme at $(j\Delta x, n\Delta t)$ provides the following expressions:

$$h_j^{n+1} = h_j^n + \Delta t\frac{\partial h}{\partial t} + \frac{\Delta t^2}{2}\frac{\partial^2 h}{\partial t^2} + \frac{\Delta t^3}{6}\frac{\partial^3 h}{\partial t^3} + h.o.t$$

$$h_{j-2}^n = h_j^n - 2\Delta x\frac{\partial h}{\partial x} + \frac{(2\Delta x)^2}{2}\frac{\partial^2 h}{\partial x^2} - \frac{(2\Delta x)^3}{6}\frac{\partial^3 h}{\partial x^3} + h.o.t$$

$$h_{j-1}^n = h_j^n - \Delta x\frac{\partial h}{\partial x} + \frac{\Delta x^2}{2}\frac{\partial^2 h}{\partial x^2} - \frac{\Delta x^3}{6}\frac{\partial^3 h}{\partial x^3} + h.o.t$$

$$h_{j+1}^n = h_j^n + \Delta x\frac{\partial h}{\partial x} + \frac{\Delta x^2}{2}\frac{\partial^2 h}{\partial x^2} + \frac{\Delta x^3}{6}\frac{\partial^3 h}{\partial x^3} + h.o.t$$

$$(4.5)$$

where 'h.o.t' stands for higher-order terms in the Taylor series expansion.

The terms in (4.5) can be substituted into (4.4) to obtain

$$h_j^n + \Delta t \frac{\partial h}{\partial t} + \frac{\Delta t^2}{2} \frac{\partial^2 h}{\partial t^2} + \frac{\Delta t^3}{6} \frac{\partial^3 h}{\partial t^3} = 0.12287[h_j^n - 2\Delta x \frac{\partial h}{\partial x} + \frac{(2\Delta x)^2}{2} \frac{\partial^2 h}{\partial x^2} - \frac{(2\Delta x)^3}{6} \frac{\partial^3 h}{\partial x^3}]$$

$$+ 0.291195[h_j^n - \Delta x \frac{\partial h}{\partial x} + \frac{\Delta x^2}{2} \frac{\partial^2 h}{\partial x^2} - \frac{\Delta x^3}{6} \frac{\partial^3 h}{\partial x^3}]$$

$$+ 0.823116[h_j^n]$$ (4.6)

$$- 0.237183[h_j^n + \Delta x \frac{\partial h}{\partial x} + \frac{\Delta x^2}{2} \frac{\partial^2 h}{\partial x^2} + \frac{\Delta x^3}{6} \frac{\partial^3 h}{\partial x^3}]$$

$$+ h.o.t$$

Rearranging the terms in (4.6) and dividing by Δt then leads to:

$$\frac{\partial h}{\partial t} + \frac{\Delta t}{2} \frac{\partial^2 h}{\partial t^2} + \frac{\Delta t^2}{6} \frac{\partial^3 h}{\partial t^3} = -0.774118 \frac{\Delta x}{\Delta t} \frac{\partial h}{\partial x} + 0.585492 \frac{\Delta x^2}{2\Delta t} \frac{\partial^2 h}{\partial x^2}$$

$$- 1.511338 \frac{\Delta x^3}{6\Delta t} \frac{\partial^3 h}{\partial x^3} - 0.000002 h_j^n + h.o.t$$ (4.7)

However, since the value $0.000002\ h_j$ is so very much smaller than the other terms, the fourth term in the right hand side of (4.7) can be neglected and the equation could be further simplified as follows:

$$\frac{\partial h}{\partial t} + (0.774118 \frac{\Delta x}{\Delta t}) \frac{\partial h}{\partial x} - (0.585492 \frac{\Delta x^2}{2\Delta t}) \frac{\partial^2 h}{\partial x^2} + (\frac{\Delta t}{2}) \frac{\partial^2 h}{\partial t^2} + (1.511338 \frac{\Delta x^3}{6\Delta t}) \frac{\partial^3 h}{\partial x^3}$$

$$+ (\frac{\Delta t^2}{6}) \frac{\partial^3 h}{\partial t^3} = h.o.t$$ (4.8)

Now by differentiating (4.2) once with respect to x and once with respect to t and then subtract the two resulting expressions in order to cancel out the cross derivative terms, and then neglect the terms of fourth-order and above, one can arrive at the following order relation:

$$\frac{\partial^2 h}{\partial t^2} \approx C^2 \frac{\partial^2 h}{\partial x^2} - 2D \frac{\partial^3 h}{\partial x^2 \partial t}$$ (4.9)

Differentiating both sides of (4.9) with respect to t and once again neglecting higher order terms gives the following expression:

$$\frac{\partial^3 h}{\partial t^3} \approx C^2 \frac{\partial^3 h}{\partial x^2 \partial t} \tag{4.10}$$

With a similar manipulation of the above equations, the following expression is also obtained:

$$\frac{\partial^3 h}{\partial x^3} \approx -\frac{1}{C}\frac{\partial^3 h}{\partial x^2 \partial t} \tag{4.11}$$

Once again, these approximations are not entirely consistent with both equations (4.1) and (4.3), but the error in the tests employed here was again found to be negligible. Substituting (4.9)-(4.11) into (4.8) and rearranging the resulting terms yields the expression:

$$\frac{\partial h}{\partial t} + (0.774118\frac{\Delta x}{\Delta t})\frac{\partial h}{\partial x} + (\frac{\Delta t}{2}C^2 - 0.585492\frac{\Delta x^2}{2\Delta t})\frac{\partial^2 h}{\partial x^2}$$
$$+ (\frac{\Delta t^2}{6}C^2 - 1.511338\frac{\Delta x^3}{6\Delta t}\frac{1}{C} - \Delta t D)\frac{\partial^3 h}{\partial x^2 \partial t} = +h.o.t \tag{4.12}$$

In this case, the C and D, which are the coefficient for the second and third terms of (4.2), can be replaced by:

$$D = (\frac{\Delta t}{2}C^2 - 0.585492\frac{\Delta x^2}{2\Delta t}) \qquad and$$

$$C = 0.774118\frac{\Delta x}{\Delta t}$$

Substituting (4.13) in (4.12) and further simplifying the equation provides the expression:

$$\frac{\partial h}{\partial t} + (0.774118\frac{\Delta x}{\Delta t})\frac{\partial h}{\partial x} + (0.006877\frac{\Delta x^2}{\Delta t})\frac{\partial^2 h}{\partial x^2} - (0.232389\,\Delta x^2)\frac{\partial^3 h}{\partial x^2 \partial t} = h.o.t \tag{4.13}$$

Due to the relatively small influence of the diffusion term in comparison to C and E in *(4.13)*, the third term in this equation can be dropped and the equation can further be simplified to:

$$\frac{\partial h}{\partial t} + (0.77411\frac{\Delta x}{\Delta t})\frac{\partial h}{\partial x} - (0.232389\,\Delta x^2)\frac{\partial^3 h}{\partial x^2 \partial t} = h.o.t \qquad (4.14)$$

Comparing (4.14) with (4.2), it is seen that the relationship that has been learned by the ANN is in fact a Boussinesq-like equation with the coefficients C and E given by the expressions:

$$C = 0.774118\frac{\Delta x}{\Delta t}$$
$$E = 0.232389\Delta x^2$$

A similar investigation was carried out using a multi-layer perceptron with one hidden layer containing four hidden neurons and linear transfer functions both in the hidden and in the output layer (as shown in Fig. 4.3.). This ANN was then trained on the same data set used above and the back propagation algorithm was found to converge to provide a residual RMS error of less than 0.002 after a relatively longer period of training time. The corresponding distribution of weights is shown in Figure 4.3.

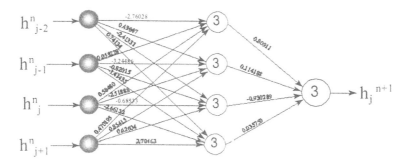

Figure 4.3. Weight distribution in Multi-layer Perceptrons with one hidden layer and a linear transfer function.

This configuration can be expressed as:

$$\begin{aligned}
h_j^{n+1} = &\ 0.80911[-2.7603h_{j-2}^n + 0.0532h_{j-1}^n + 0.5848h_j^n + 0.4702h_{j+1}^n] \\
&+ 0.11419[0.4385h_{j-2}^n - 3.2449h_{j-1}^n - 2.5189h_j^n + 0.8541h_{j+1}^n] \\
&- 0.93029[-2.4133h_{j-2}^n - 0.8202h_{j-1}^n - 0.6853h_j^n + 0.6260h_{j+1}^n] \\
&+ 0.03576[0.7413h_{j-2}^n - 1.4144h_{j-1}^n - 2.6613h_j^n - 2.704633h_{j+1}^n]
\end{aligned} \qquad (4.15)$$

which can be further simplified to:

$$h_j^{n+1} = 0.088273h_{j-2}^n + 0.384945h_{j-1}^n + 0.727927h_j^n - 0.201145h_{j+1}^n \qquad (4.16)$$

Although the coefficients in (4.16) are somewhat different from those of (4.4), if each
variable in this equation is replaced with the corresponding Taylor series approximation
and the resulting expression simplified by neglecting the higher order terms, as outlined in
the case of the two-layer network, the following expression can be obtained:

$$\frac{\partial h}{\partial t} + (0.76264\frac{\Delta x}{\Delta t})\frac{\partial h}{\partial x} + (0.02236\frac{\Delta x^2}{\Delta t})\frac{\partial^2 h}{\partial x^2} - (0.20784\,\Delta x^2)\frac{\partial^3 h}{\partial x^2 \partial t} = h.o.t \qquad (4.17)$$

This expression in (4.17) is once again similar to (4.2), but with slightly different
coefficients. Although these coefficients, in general, are found to be similar to those of the
two-layered net, the analysis gives a relatively higher diffusion effect and lesser values of
the celerity and the coefficient of dispersion. This is most probably due to the relatively
larger value of root mean square error (RMSE) that could be reached during the training
time that was practically acceptable. Therefore, the remaining analysis is mainly focused
on the results obtained by the training of the two-layered networks.

Since the celerities of the waves in a channel vary with the depth of flow, it was required
to repeat the experiment for different values of still-water depths. As the ratio h/L_0 is less
than 0.22 for the average depths considered, the classical Boussinesq equation in the
MIKE21 wave module was again used to generate the training data. ANN training and
analyses similar to the one shown above were carried out on the flow data using two-
layered linear ANNs and a summary of the result is presented in Table 4.1.

Table 4.1. Summary of coefficients of the Boussinesq equation for different water depths.

Depth of flow	α	C	β	E
3m	0.54698	5.4698	0.06386	1.5966
4m	0.64505	6.4505	0.11539	2.8848
5m	0.71665	7.1665	0.17862	4.4654
6m	0.77411	7.7411	0.23240	5.8100
7m	0.83104	8.3104	0.30355	7.5888
8m	0.87627	8.7627	0.34786	8.6965

In this table the coefficient for the terms in the Equation (4.2) are given by:

$$C = \alpha \frac{\Delta x}{\Delta t}, \qquad E = \beta \Delta x^2$$

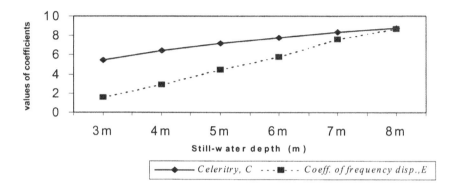

Figure 4.4. Variation in the values of the coefficient of celerity and frequency dispertion with depth.

The variation in the value of these coefficients with the average depth of flow is shown in Figure 4.4. From this figure, it is seen that both the celerity and coefficient of dispersion increase with the increase in the average depth of flow, as could be expected.

The physical celerity for the waves corresponding to the different still water depths is estimated to be of the order of $(gh_0)^{1/2}$, where g is the gravity, and h_0 is the average depth of flow (still-water depth). From equation (4.1a) one can see that the frequency dispersion factor E is on the order of $h_0^2/6$. In fact the values so obtained have been compared with their corresponding values found after analysing the weights of the ANNs and the results are presented in Table 4.2. and plotted in Figures 4.5. and 4.6.

Table 4.2. Comparisons of observed versus calculated celerity and frequency dispersion factor for different depths of flow.

Average depth of flow	3m	4m	5m	6m	7m	8m
Calculated physical celerity $(gh_0)^{1/2}$ (m/sec)	5.4249	6.2642	7.004	7.6720	8.2867	8.8589
Celerity, C, from ANN, (m/sec)	5.4698	6.4505	7.1665	7.7411	8.3104	8.7627
Frequency dispersion factor, $h_0^2/6$ (m^2)	1.5000	2.6667	4.1667	6.0000	8.1667	10.667
Dispersion factor, E, from ANN, (m^2)	1.5966	2.8848	4.4654	5.8100	7.5888	8.6965

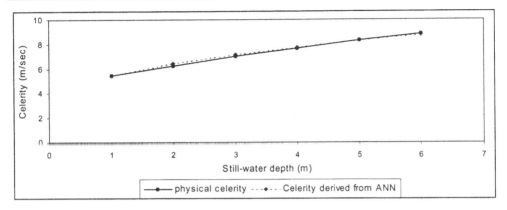

Figure 4.5. Comparison of the celerity C obtained from the ANN with the physical phase celerity $(gh_0)^{1/2}$.

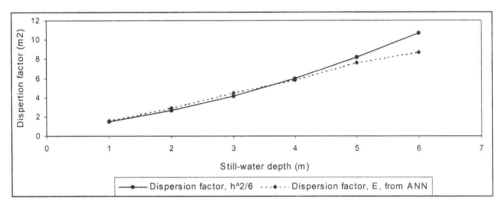

Figure 4.6. Comparison of the frequency dispertion factor E obtained from the ANN with the actual value of $h_0^2/6$.

From the plot in Figure 4.5., it is seen that the ANN gives celerity values which are very close to the physical phase celerity. Accordingly, it seems that the ANN suffices to represent the celerity of the Boussinesq wave to a reasonable accuracy for the range of flow depth considered in this investigation, just as it had done in the case of waves of constant form.

In a similar way, Figure 4.6. shows that the values of the dispersion coefficient E for the different depths of flow is quite close to the term $h_0^2/6$ for the relatively smaller depths of flow (3m to 6m). On the other hand, with an increase in depth beyond 6m, the E values derived from the ANN start to depart from those provided by the coefficient of the third order term in equation (4.1a). One reason for this may be that the approximations introduced during the analysis lead to the form of the Boussinesq equation becoming less accurate with the increase in the depth of flow.

4.4   The Encapsulation of Numerical-Hydraulic Models in Artificial Neural Networks

The practice of the numerical simulation of flows and other processes occurring in water has now matured into an established and efficient part of hydraulics: numerical-hydraulic models have become reliable tools for the analysis, design and management of a wide range of water-based assets. At the same time, however, these models have often become very extended: coastal and dam break models with hundreds of thousand, or even of the order of a million grid points or urban drainage system models with a similar number of nodes are becoming increasingly common, and these models make heavy demands on computing capacity and time. Although these may be acceptable in many situations, such as in design, planning and many management operations, there are a number of applications for which very long computing times are unacceptable. As explained in the introductory chapter, the most significant of these at the present time is that of the real-time control (RTC) of such assets as rivers and urban drainage systems where only a relatively short time is often available between a forecast of expected inputs and the need to set counter-measures in action to alleviate the effects of flooding and other disembursements.

Given this divergence between the computational-time requirements of numerical models (usually further exacerbated by the need to run a considerable number of simulations when making-up the optimal control procedures and the need to constantly update the control procedures as the flood event develops in time) the need arises to reduce the time needed to simulate the impact of rain events and similar natural interventions on hydraulic systems. The most obvious way to do this without significant loss of accuracy is to encapsulate the numerical model in a form that will allow a much more rapid response in terms of outputs to any of a wide range of given inputs (e.g. Masood-UI-Hassan et al., 1995). It has been demonstrated explicitly in з 4.1 how domain knowledge encapsulated in numerical models can in turn be encapsulated in artificial neural networks. In more general hydroinformatics terms, this involves the replacement of one sign vehicle - providing the form of the numerical model - by another sign vehicle - which in the present case has the form of the weights in an artificial neural network of one kind or the other.

The applicability of artificial neural networks for encapsulating a numerical hydraulic model to simulate water resources systems has been already investigated for a number of cases. One such study was the simulation of flows in Apure river basin in Venezuela (Solomatine and Torres, 1996; Dibike and Solomatine, 2001). The results of this study showed that, if the appropriate network architecture and training algorithms are identified, an ANN has the ability to emulate a numerical-hydraulic model in relating downstream flow ordinates at a particular place to a large number of inflow ordinates at upstream locations, provided, of course, that the inputs do in fact contribute in some way to the flow at the downstream point. If upstream events are within the range of values experienced by the network during the training session, ANN models can be used to forecast the downstream discharges given any set of upstream inflows. The results of the above case study indicate that, when it is required to incorporate a modelling tool which can simulate a river system within an optimisation loop, these type of simplified ANN models could be employed to emulate numerical-hydraulic models with acceptable accuracy. Since a large number of simulation runs is often required to find the optimal solution for a problem, the use of the original, more sophisticated, models is often too time-consuming to be practical. Hence ANN emulation provide a viable engineering alternative.

In a similar way, another study (Maskey et al., 2000) also demonstrated the potential applicability of ANN models in selecting an optimal pumping strategy for the remediation of a contaminated aquifer. In this case, ANNs were trained to emulate the physically based groundwater models MODFLOW and MODPATH. First the networks were trained using the data generated by these models. The result presented in Figure 4.7. showed that the ANN model was able to give reasonably good approximations on the optimal aquifer cleanup time. The resulting ANN models were then coupled with a global optimisation tool, GLOBE (Solomatine, 1999), to establish optimal pumping strategies for the contaminated plume removal.

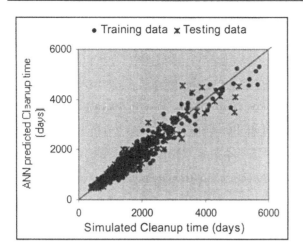

Figures 4.7. Scatter plots of network training and validation for cleanup time with 3 pumps.

Both the physically-based and ANN models were coupled with the optimisation tool (GLOBE) which contains three different global optimisation algorithms, namely a genetic algorithm, an adaptive cluster covering and a controlled random search (see Solomatine, 1999). The results in Figures 4.8 (a & b)., which correspond to three pumping wells, show that the ANN model coupled with the optimization tool was able to provide a reasonably good approximation to the optimal aquifer cleanup time while requiring a very short computational time (being, on average, 17 times faster than the physically based model) to arrive at approximately the same solution. Moreover, the simplicity of the ANN model, both in its use and for coupling with any global optimisation tool, was yet another advantage.

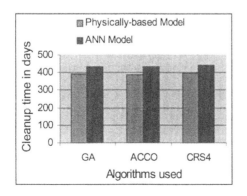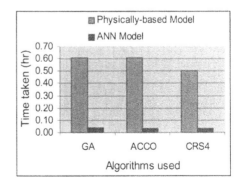

Figure 4.8(a & b): Optimal solutions for (a) cleanup time and (b) running time required using physically based and ANN models simulations for 3 wells.

In terms of the optimal solutions, all three algorithms produced more or less similar optimal cleanup times, while the running time taken to compute the optimal solution by the different algorithms are only slightly different, but are all considerably faster than the original physically-based model.

4.5   Discussion

Information processing can generally be categorised into symbolic paradigms (exemplified by e.g. rules or partial differential equations), and connectionist descriptions (exemplified by e.g. ANNs). The symbolic paradigm (using the 'mathematical language') is very well suited to structural knowledge representation in a way quite comprehensible to human beings (provided they speak the language of mathematics). On the other hand, the sub-symbolic connectionist paradigm is often superior for *learning* with *associative memory* and *recall,* in a way similar to the human brain. Because its basic mechanism is that of a similarity-based inference, information processing with ANNs is often more flexible and robust in the presence of noise. Moreover, since it can represent physical signals as well, the connectionist paradigm seems more compatible with the physical world than is the symbolic paradigm (Cloete, 2000).

The preceding numerical experiments and the corresponding results indicate that in the case that the mathematical formulation of a physical process is known, - even if it is implicit and distributed in the weights -, the ANNs can be shown to encapsulate the same knowledge, or exhibit the same semantic content over a useful range of applications, as do these continuum equations. This is demonstrated by actually restoring the equations from the weights of the ANNs. Previous research (Minns, 1998) has also shown that, in the simple case of pure advection with constant velocity, a linear ANN is capable of learning the exact solution, which is also exactly equivalent to the differential equation description. Promising results were also obtained during this study for the case of the propagation of short-period waves. The investigation showed that, by analysing the weights obtained by training the linear ANNs, it was possible to derive the governing one-dimensional form of the Boussinesq-like partial differential equation. The coefficients of celerity in the derived equations corresponding to the different still water depths were found to match very well with the physical celerity for each flow condition over a wide range of still-water depths.

The analysis also shows acceptably accurate values and trend for the coefficients of frequency dispersion, albeit over a somewhat narrower range. Experience in the present work suggests that the use of more time steps, the use of longer ANN learning periods and a more accurate and correspondingly more detailed treatment of the elimination processes in the Taylor series analysis would all contribute to still further improving the results obtained so far.

The brief overview of case studies on the encapsulation of numerical-hydraulic models in artificial neural networks also suggests the future direction of these techniques as means for emulating numerical algorithms describing a particular water system by learning from data generated with a more elaborate numerical model. These ANN models possess the extra advantage that they are very fast and very convenient in cases where a large number of simulations has to be done in a relatively short time, e.g. to find the optimum operation policy for real time control of water resource systems.

Chapter 5

Simulation of Hydrodynamic Processes Using Artificial Neural Networks

5.1   Introduction

The practice of numerical simulation of flows and other processes occurring in water, based on the formulation and solution of mathematical relationships expressing known hydraulic principles, was once the main area of research in computational hydraulics and has now matured into an established and important part of hydraulics generally. At the same time, hydroinformatics has begun to become a dominant paradigm in several of the research institutes and organisations that are major suppliers of the latest generations of modelling systems. The hydroinformatics paradigm differs from the earlier computational hydraulics paradigm in a number of ways. One of these differences is that, whereas computational hydraulics used data mainly to calibrate and validate models that were already largely preconceived in the minds and papers of the computational hydraulicians, hydroinformatics increasingly uses data to evolve the models themselves, as 'emergent phenomena', using such methods of *metamodelling* as artificial neural networks, evolutionary algorithms, etc. It is accordingly essential to explore at least some of the ways in which the developments in such 'soft' computing techniques that have more recently been introduced into hydroinformatics might also be applied to the design and construction of a new generation of computational hydraulic engines (Abbott, 1997a).

The central dogma in this case may be stated as follows:

> To the extent that the new computational engine learns to reproduce the behaviour of a
> model that is determined by the set of beliefs that this model encapsulates, then the engine
> encapsulates these same beliefs.

Thus the new engine need not incorporate these beliefs 'explicitly', but may learn them from existing (and often much more complicated) computational hydraulic models. The most obvious way to do this for prediction purposes is through the training of an ANN to reproduce the behaviour at one place at one time from the behaviour at other places at earlier times. In the event that such learning proceeds on the basis of the results provided by a numerical model, the places and times can be related to the 'grid points' in the numerical model (Babovic, 1997). The objective of this work is, therefore, to investigate the possibility of using systems composed of agents consisting only of ANNs as modelling tools, and in this particular case, for the simulation of a tidal flow in a two-dimensional flow field. In this particular case, this involves the modelling of a process that evolves in time so that the ANNs themselves function as non-linear dynamic systems that effectively model one or more mutually-dependent time series. Different types of ANN architectures are investigated in order to asses their ability and relative performance in encapsulating the site-specific knowledge and data necessary to reproduce the temporal sequence of states observed in the modelled area.

5.2   Two Dimensional (2D) Hydrodynamic Modelling

In this study, a water circulation model of Donegal Bay in Ireland (see Fig. 1.) has been considered. The model was set up using the MIKE 21 modelling system (MIKE 21 BW, 1996). The hydrodynamic model in MIKE 21 is a general numerical modelling system for the simulation of water levels and flows in estuaries, bays and coastal areas. It simulates unsteady two-dimensional flows in one layer (vertically integrated velocity) fluids and has been applied in a very large number of studies (Abbott, 1997b). The following equations, corresponding to the conservation of mass and momentum integrated over the vertical, describe the flow and water level variations:

Continuity

$$\frac{\partial \zeta}{\partial t} + \frac{\partial p}{\partial x} + \frac{\partial q}{\partial y} = 0 \tag{5.1}$$

x-momentum

$$\frac{\partial p}{\partial t} + \frac{\partial}{\partial x}\left(\frac{p^2}{h}\right) + \frac{\partial}{\partial y}\left(\frac{pq}{h}\right) + gh\frac{\partial \zeta}{\partial x} + \frac{gp\sqrt{p^2+q^2}}{C^2 h^2} - \frac{1}{\rho_w}\left[\frac{\partial}{\partial x}\left(h\tau_{xx}\right) + \frac{\partial}{\partial y}\left(h\tau_{xy}\right)\right]$$
$$-\Omega q - fVV_x + \frac{h}{\rho_w}\frac{\partial}{\partial x}\left(p_a\right) = 0 \tag{5.2}$$

y-momentum

$$\frac{\partial p}{\partial t} + \frac{\partial}{\partial y}\left(\frac{q^2}{h}\right) + \frac{\partial}{\partial x}\left(\frac{pq}{h}\right) + gh\frac{\partial \zeta}{\partial y} + \frac{gp\sqrt{p^2+q^2}}{C^2 h^2} - \frac{1}{\rho_w}\left[\frac{\partial}{\partial y}\left(h\tau_{yy}\right) + \frac{\partial}{\partial x}\left(h\tau_{xy}\right)\right]$$
$$-\Omega p - fVV_y + \frac{h}{\rho_w}\frac{\partial}{\partial y}\left(p_a\right) = 0 \tag{5.3}$$

where:

$h(x,y,t)$	-water depth (m)
$\zeta(x,y,t)$	-surface elevation (m)
$p,q,(x,y,t)$	-flux density in x- and y- directions ($=$ uh and vh respectively, (m^3/s/m))
$u, v\ (x,y,t)$	-depth averaged velocity in x and y direction (m/s)
$C(x,y)$	-Chezy resistance ($m^{1/2}$/s)
g	-acceleration due to gravity (m/s^2)
$f(V)$	-wind friction factor
$V, V_x, V_y, (x,y,t)$	-wind speed and components in the x- and y- direction (m/s)
$\Omega(x,y)$	-Coriolis Parameter, latitude dependent
pa (x,y,t)	-atmospheric pressure (kg/ms^2)
ρ_w	-density of water (kg/m^3)
x,y	-space co-ordinates (m)
t	-time co-ordinate (s)
$\tau_{xx}, \tau_{xy}, \tau_{yy}$	-components of the effective shear stress

The MIKE 21 hydrodynamic modelling system solves these equations along with the initial and boundary conditions for a specific problem area. It makes use of the so-called Alternating Direction Implicit (ADI) technique to integrate the equations for mass and momentum conservation in the space-time domain (Abbott and Minns, 1998). The equation matrices that result for each direction and each individual grid line are resolved by using a Double Sweep (DS) algorithm.

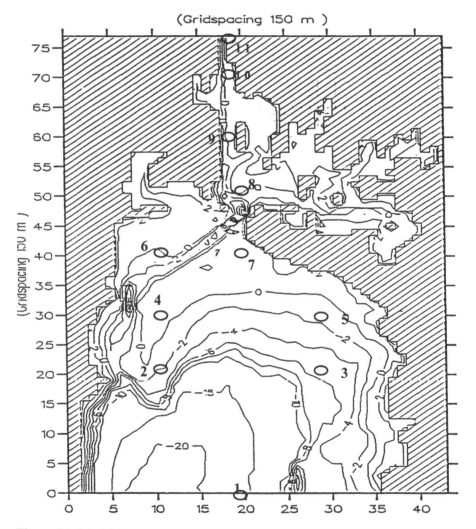

Figure 5.1. Model bathymetry of Donegal Bay and the positions identified for the ANN investigation.

In the case of Donegal Bay, a grid spacing of 150 meter was chosen as this was the maximum size that could be used to define the tidal channels in the estuary. The time step was chosen to be 60 seconds since it results in a maximum Courant number of 5.9 in the deepest part of the model and in Courant numbers of 2 to 3 in the estuary. Simultaneous measurements of the water level close to the open boundary and near Donegal had been made and the hydrodynamic (HD) model was accordingly calibrated so that the simulated and measured water level variation near Donegal showed satisfactory agreement.

During the simulation experiment, the water level at the open boundary (incoming tide) was prescribed as a sine wave with time-varying tidal heights and covering different ranges of free-surface water levels. Tidal waves with amplitudes of 3.5m (corresponding to a spring tide), 1.5m (corresponding to a neap tide) and an intermediate amplitude of 2.5m, each having a period of 12.5 hours, were considered, as shown in Figure 5.2. The HD model was then applied to generate the values of water level and velocity at each grid point in the model area. These data were then transformed to time series of depths and velocities at the different points in the model area that were identified with the ANN-based agents. The HD simulation was done for each range of tides for a period of about two days (four tidal cycles) so that there would be enough data for the training and verification of the networks.

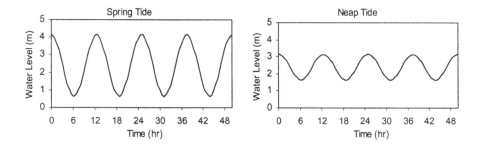

Figure 5.2. Time series of input water levels at the model's open boundary.

5.3   Simulation of Water Levels and Velocities Using ANNs

Representative grid points for the location of the ANN-based agents were identified in the M21-based numerical model area and the time series of water levels and velocities at these points and at the open boundary were prepared. An important prerequisite for employing ANNs is the provision of a representative set of input - output data and the relevance of the input to the output. Therefore, it is necessary to understand the underlying physics of a system before determining the input-output data used for ANN modelling. In hydrodynamic problems, information is transported in the space time domain along the dynamic 'long wave' characteristics. These characteristics bound the so-called 'domain of dependency' of a point in the space-time domain (see Fig 5.3.). Only information within a point's domain of dependency can affect that point's state. The final goal in this exercise is to present the ANN with input containing the maximum possible information content with respect to the output which will assist the network in creating a better relationship. In the case of a two-dimensional flow this situation is further complicated by the diffusion of influence over the entire domain of dependence of the point in question (Abbott, 1966)

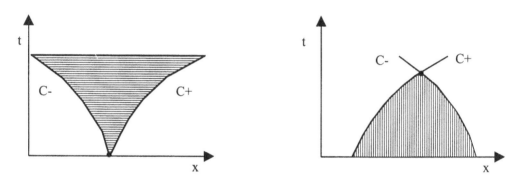

Figure 5.3. Propagation of a disturbance and the domain of dependency in channel flow [adapted from Cunge et al., 1980].

5.3.1 Simulation with boundary data as input to the ANN-based agents

In the first set of experiments, the time series of water levels at the open boundary of the model was considered as the only input to all ANN-based agents, while each of the corresponding time series of water levels and velocities at the different internal grid points was considered as output from these agents. As the variation of the water level with time was relatively small, the data was prepared at 10 minutes intervals and it was divided into training, cross validation and testing periods. In most of the cases, the network was trained with the data corresponding to the spring and neap tides and tested with the data corresponding to the intermediate tidal range (see Fig. 5.2.)

The first neural network investigation was made with MLP networks. However, before using the MLP network, it was necessary to pre-process the data in such a way that the input to the network again consisted not only of the boundary data at the present time, but also those at the appropriate previous time steps. The number of these previous time steps constituted what is known as the 'window length' of the influences. The optimum size of this window in this feasibility study was determined by experimenting several times with different numbers of inputs and performing a corresponding sensitivity analysis. This process was found to be one particularly time-consuming part of the whole investigation. Figure 5.4. shows the result of this analysis for one particular experiment (at point No. 11 on Fig. 5.1.). From this experiment it was found that the optimum window length was proportional to the distance of the point where the water level was required to be simulated from the boundary that constitutes the input, a result consistent with the classical theory of characteristic varieties (Abbott, 1966).

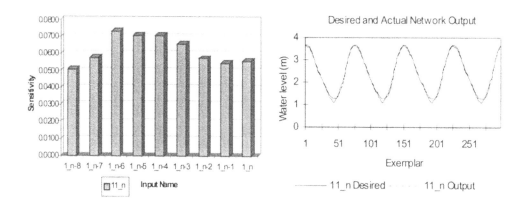

Figure 5.4. Sensitivity of the network for inputs with different lag times and the network performance with the optimum window length.

After the optimum window length was determined, the MLP network was found to converge in a relatively smaller number of iterations. However each iteration took quite a considerable time in this case because of the relatively large number (13 in this particular case) of hidden neurons required for the training. The resulting trained network was however then found to fit quite well, as schematized in Figure 5.4., with a mean square error of 0.0045. Similar experiments were also made with multiple-output networks where the network was trained to relate the series of water level inputs at the model boundary with the water levels of 3 to 7 different points in the model area, and satisfactory results were again obtained.

The other type of ANN employed in these exploratory studies was the time-delay neural network (TDNN). The use of this network, as explained in ɔ 2.4, avoided the need to determine the size of the temporal window explicitly. It was also found to require a relatively smaller number of neurons in the hidden layer. Each iteration was fast, although a relatively larger number of iterations was required for convergence during training. In this particular case, the velocities in the x and y directions were specified as output in addition to the water levels at the points identified within the model area. Figure 5.5. shows the performance of the trained network on the test data at one representative point in the model area (namely point No. 8 in Fig. 1.) where the network output fitted the numerical output with MSE values of 0.005, 0.0008 and 0.0012 for h, u and v respectively. Even better performances were observed at some other points that were associated with a relatively smoother velocity field.

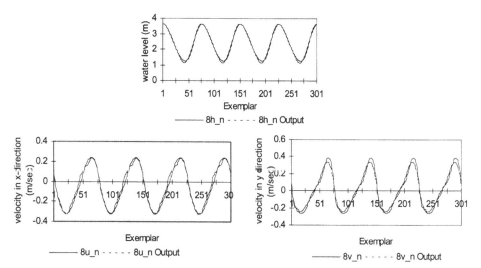

Figure 5.5. Performance of a recurrent network with feedback from the input layer to estimate the water level and the velocities in the x and y directions at point No. 8 in the model.

Similar investigations were also made by considering the absolute magnitude and the corresponding direction instead of the two perpendicular components of the velocity vector as an output from the network. The direction of the velocity vector was represented in terms of radians, and so varying between $-\pi$ and π. A typical performance of the network is shown in Figure 5.6. Here it can be observed that the network performed well in representing the water level and the magnitude of the velocity but had a certain local difficulty in representing the direction of the velocity correctly. A closer examination of the problem revealed that this situation arose due to the circular nature of the function that was to be approximated. This means that since the direction of velocity takes a value between $-\pi$ and π (or possibly between 0 and 2π), the network had difficulty in representing the situation around the extreme angles and usually tended to give the average of the possible outcomes.

Investigations made with the Jordan Network also showed quite similar characteristics to those described above. In some cases this network was found to perform better when the number of units in the output layer was increased, corresponding to a larger number of point in the flow field being considered as an output for a given input boundary condition during the simulation. A possible reason for this is that, in addition to the boundary data, not only the output of the network at some previous time steps of a given point, but also the output of the surrounding points was taken as input to the network. Figure 5.7. shows the performance of the trained network at point 8, where the network output fits the actual one with MSEs of 0.033, 0.0011 and 0.0013 for h, u and v respectively.

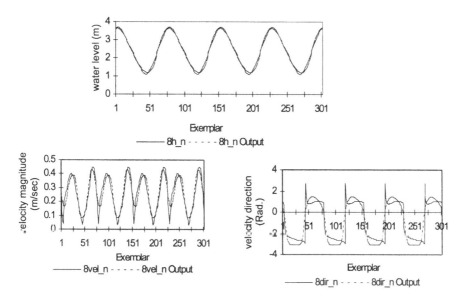

Figure 5.6. Performance of the recurrent network with feedback from the input layer to estimate the water level and the velocities vector at point No. 8 in the model.

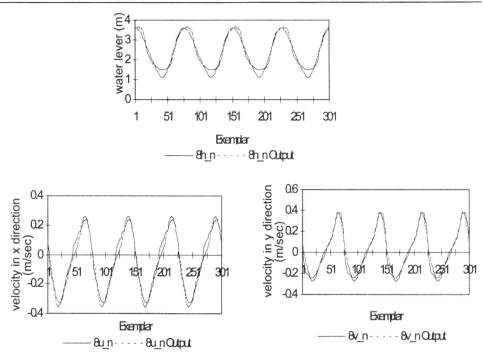

Figure 5.7. Performance of the recurrent network with feedback from the output layer to estimate the water level and the velocities in the x and y directions at point No. 8 in the model.

5.3.2 ANN-based agents as message passing structures

Despite the excellence of fit obtained using the configuration described in ϶ 5.3.1, a second set of experiment was performed in order to check whether using ANN-agents as message passing structures on an unstructured grid could contribute to the accuracy of the simulation results. This time, the ANN-agents at each selected grid point received their inputs from the present and previous values of the velocities and water levels obtaining at the nearest surrounding agents, which in turn could be associated with certain grid points. The agents then predicted the corresponding values of the variables at that point for the next time step. This also meant that water levels and velocities at the present and previous time steps of each selected grid point were passed as inputs to the ANN agents of the nearest grid points to predict the water levels and velocities at the next time step at those points. A representative part of the message-passing structure used for the model area considered in this study is shown in Figure 5.8. The data for this investigation was prepared at 3 minute intervals based on the average distance between the grid points selected for the message-passing structure. Each such structure that was associated with a given agent was then a measure of the domain of responsibility of other agents towards that agent (Abbott, 1997a).

During the simulation experiments, messages containing information about the present and past states of the flow variables from ANN-agents at each grid point considered in the model area were passed to the ANN-agents at the nearest surrounding 'grid points'. At the same time, each ANN-agent received similar information from the ANN-agents situated within its nearest neighborhood, as already schematized in Figure 5.8.

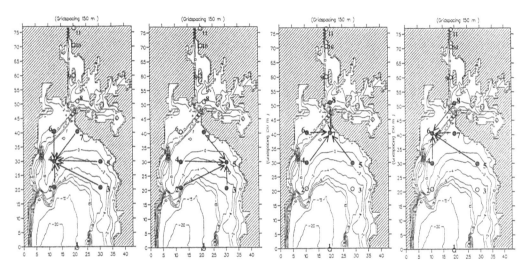

Figure 5.8. Message passing structures of ANN-agents with a non structured grid (information about the previous state of the flow variables being received from nearby surrounding grid points).

After some preliminary investigations, the time-delay neural network (TDNN) was found to be the most suitable architecture and it was typically used with a delay of three time steps in the input layer. This number of time steps to be delayed was, of course, determined on the basis of its distances along the predominant propagation paths of an ANN-agent from the surrounding grid points from which information was passing, and also of course the time difference between consecutive discrete points in the data. These numbers could again be determined automatically by providing the ANN agent with data extending back in time and allowing the ANN to select its own weights in time. Typical verification results of this investigation are shown in Figure 5.9. for two grid points in the model area. In the verification period, by way of examples, the network output fitted the simulated one with an MSE for *h*, *u* and *v* respectively of 0.0065, 0.0001 and 0.0017 at point 7 and 0.004, 0.0007 and 0.0005 at point 8.

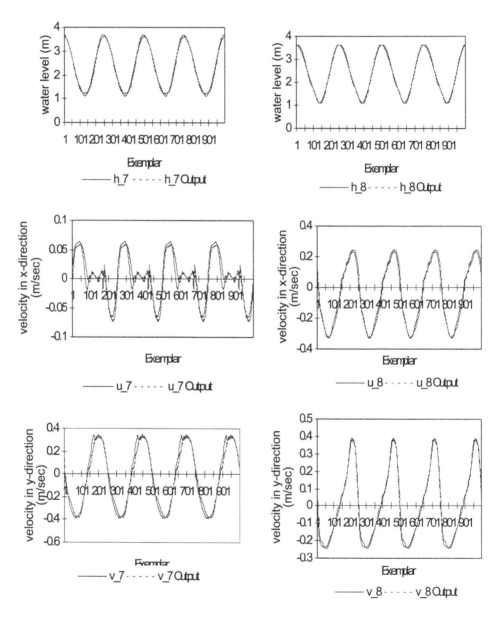

Figure 5.9. Typical verification performance of the TDNN (at point 7 and 8) in simulating the flow variables (water level and velocities) at the present time using data taken from the surrounding grid points taken at earlier times.

5.4   Discussion

The objective of this study was to investigate in a preliminary way the feasibility of using agents composed almost exclusively of ANNs for hydrodynamic modelling in two-dimensions. For this purpose, different ANN architectures were investigated within different settings of the problem. The main feature of the time series of the flow variables (velocities and water level) at any grid point in the flow field was that the successive observations were highly interdependent and the future values of the time series at a given point could be predicted from the present and past time series values of these and the surrounding grid points in the flow field.

In the first set of experiment the time series of water levels at the open boundary of the two-dimensional numerical model was used as the only input to eleven ANNs-based agents, corresponding to eleven different internal 'grid points'. The recurrent networks with feed-back, either from the output layer or from the input layer (with input delay), demonstrated a satisfactory performance. This could however be because the model area considered was not particularly demanding on accuracy, even though it sufficed for this purely exploratory study. The applicability of the absolute magnitude and the corresponding direction of velocities in polar coordinate instead of their components in Cartesian coordinates were also investigated within this context. However, this practice was found to reduce the apparent accuracy of the network due to ambiguities in the circular nature of the representation of the direction.

In the second set of experiments, where the ANN-agents were distributed in the model area with a non- structured grid, each ANN-agent received its information (input) from the time series of flow variables (output) from the other ANN-agents in the nearest surrounding grid points. The TDNN was then used, with a uniform delay of three time steps, in order to predict the flow variables for the next time step. This resulted in a good performance in predicting the water levels and the velocity components in the x and y directions. Therefore, once we trained ANN-agents at all points of interest in a given water system, then these ANN-agents can be integrated in to a single system that can simulate the hydrodynamic variables especially at those points of interest. Such approached is therefore believed to significantly reduce the computational time and resources required for simulation as it avoids calculation of the values of the flow variables at every grid point as in the case of conventional numerical models.

Chapter 6

Developing Generic Hydrodynamic Models Using Artificial Neural Networks

6.1 Introduction

The problem of determining a fluid flow is usually divided into two stages. The first of these is concerned with a description of the flow of the fluid in such general terms that this description will hold at each and every point in the domain of the solution at all times. Such a description is said to be generic to the class of flows concerned. The result is either a so-called 'point description' such as a partial differential equation, or an 'interval description', such as an integral equation. The second stage of the problem is concerned with transforming this 'point' or 'interval' representation into a representation that is distributed over the entire domain of the solution at all times, such as is for example realised by the process of integrating a partial differential equation. The difficulties experienced in integrating analytically over complicated domains has led to the widespread and now almost universal use of numerical methods, in which point and integral descriptions are extended to finite spatial descriptions that are maintained over finite time intervals, thus providing solution procedures of finite cardinality.

At the same time, the ever increasing demand for higher accuracy, higher speed, unconditional stability, ease of instantiation and other requirements have led to codes of ever greater length and complexity, such as are ever more difficult to develop and maintain by human means. The question then naturally poses itself of how numerical schemes might be generated by using other practices than those of the now 'traditional' numerical methods.

6.2   The Context of This Work

Numerical modelling is nowadays less and less an isolated activity, being increasingly integrated into other activities, such as field-measurement programs, data-assimilation facilities and automated calibration procedures. Similarly, such integrated systems are increasingly widely used to provide such services as real-time control and flood warning. All of these developments influence the specifications for modelling capabilities and especially on the side of the shorter response times that are now often required from the modelling systems.

Moreover, development in software engineering, and even in some branches of computer science, have led to the development of new methods, tools and environments that do not lend themselves readily to the existing numerical-hydraulic paradigm. These difficulties have led to the notion of moving towards new modelling paradigms, that would fit better into current practices and controlling possibilities (Abbott, 1997a). One essential component in this development is that of constructing new classes of models more or less automatically from data obtained from existing numerical-hydraulic models and / or from available field measurements, while guiding future measurement programs in the direction of serving the corresponding class of learning processes.

6.3   The Formulation of the Classes of Schemes Within the Existing
Paradigm

The objective of the numerical-hydraulic modelling of a water system is to establish the different flow variables, such as water depths and velocities, as functions of time over the area of interest such that, knowing the initial state of the system and the different forcings, the state of the system at any future time can be simulated. The conventional way of achieving this objective is first to derive relations between flow variables from physical laws used to express the conservation of certain quantities, such as mass, momentum and energy. This usually results in some system of partial differential equations which reasonably describes the flow phenomena. In the general case, these equations have no analytical solutions and have to be solved approximately by means of numerical methods, which result, as introduced earlier, in values of the solution (flow variables) being defined only at discrete points in space and time. The way in which the computation proceeds from values of dependent variables at grid points at one time level to their values at the next time level naturally depends on the computational scheme considered (Abbott and Basco, 1989; Abbott and Minns, 1998).

6.3.1   The simple explicit formulation used as a test case

Consider the primitive equations of de Saint Venant for one-dimensional nearly-horizontal, free-surface flows in the Eulerian form on a horizontal bed without resistance or other momentum-diffusive terms:

Continuity $\qquad\qquad \dfrac{\partial h}{\partial t} + h \dfrac{\partial u}{\partial x} + u \dfrac{\partial h}{\partial x} = 0$ $\qquad\qquad\qquad\qquad$ (6.1)

Momentum $\qquad\quad \dfrac{\partial u}{\partial t} + u \dfrac{\partial u}{\partial x} + g \dfrac{\partial h}{\partial x} = 0$ $\qquad\qquad\qquad\qquad$ (6.2)

Where h is the water depth and u is the depth-averaged water velocity.

For very long waves with amplitudes that are relatively small as compared with the water depth, the advective acceleration terms uu_x and the uh_x in the mass equation are of lower order than the other terms. In such cases, the primitive Equations (6.1) and (6.2) are often used in a *linearised form*:

$$\frac{\partial h'}{\partial t} + h_0 \frac{\partial u}{\partial x} = 0 \qquad\qquad\qquad\qquad\qquad (6.3)$$

$$\frac{\partial u}{\partial t} + g \frac{\partial h'}{\partial x} = 0 \qquad\qquad\qquad\qquad\qquad (6.4)$$

where $h = h_0 + h'$, with h_0 the still water depth and $h' << h_0$ or, correspondingly when the motions are determined by the variations in elevation only, $u << gh$, and $h = h_0$ in the coefficient correspondingly.

These equations can be simplified, in turn, to give the following equations in a single variable:

$$\frac{\partial^2 h'}{\partial t^2} - c^2 \frac{\partial^2 h'}{\partial x^2} = 0 \tag{6.5}$$

$$\frac{\partial^2 u}{\partial t^2} - c^2 \frac{\partial^2 u}{\partial x^2} = 0 \tag{6.6}$$

where $c^2, = gh_0$, is the square of the mean celerity

In two-dimensional flows, the continuity and momentum equations can be linearised in the same way to provide the equation:

$$\frac{\partial^2 h'}{\partial t^2} - c^2 (\frac{\partial^2 h'}{\partial x^2} + \frac{\partial^2 h'}{\partial y^2}) = 0 \tag{6.7}$$

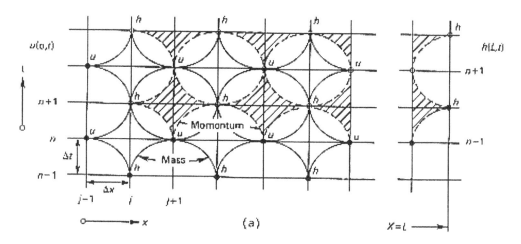

Figure 6.1. Schematisation of staggered grid (leap frog method).

Consider now a staggered-grid, three-time-level scheme, as depicted in Figure 6.1. Using centred space and time differences, the representation in finite difference form of Equations (6.3) and (6.4) becomes:

$$\frac{h'^{n+1}_j - h'^{n-1}_j}{2\Delta t} + h_0 \frac{u^n_{j+1} - u^n_{j-1}}{2\Delta x} = 0 \tag{6.8}$$

$$\frac{u_{j+1}^{n+2}-u_{j+1}^{n}}{2\Delta t}+g\frac{h_{j+2}'^{n+1}-h_{j}'^{n+1}}{2\Delta x}=0 \tag{6.9}$$

Equations (6.8) and (6.9) can be solved explicitly for h_{j}^{n+1} and u_{j+1}^{n+2}, both of which variables are staggered in space and time. Such a three level representation is usually referred to as a *leapfrog scheme* and requires initial data at time level n and n-1.

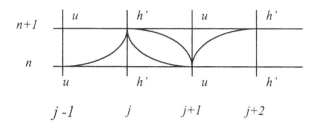

Figure 6.2. A variation of the *leap frog* scheme on a staggered grid.

A simple variation on Figure 6.1. can provide a two-level scheme as shown in Figure 6.2. In this case, initial values of u and h need to be known only at one time level. The corresponding finite difference equations can then be described by:

Continuity $$\frac{h_{j}'^{n+1}-h_{j}'^{n}}{\Delta t}+h_{0}\frac{u_{j+1}^{n}-u_{j-1}^{n}}{2\Delta x}=0 \tag{6.10}$$

Momentum $$\frac{u_{j+1}^{n+1}-u_{j+1}^{n}}{\Delta t}+g\frac{h_{j+2}'^{n+1}-h_{j}'^{n+1}}{2\Delta x}=0 \tag{6.11}$$

We can solve (6.10) and (6.11) explicitly for h and u values which are staggered in space. Equation (6.10) can be reformulated as

$$h_{j}'^{n+1}=h_{j}'^{n}-\frac{h_{0}\Delta t}{2\Delta x}(u_{j+1}^{n}-u_{j-1}^{n}) \tag{6.12}$$

Multiplying both the numerator and denominator of the second term in the right hand side of (6.12) by a gravity term g and substituting the square of the celerity, c^{2}, for (gh_{0}), (6.12) can be re-written in the following form:

$$h'^{n+1}_j = k * u^n_{j-1} + l * h'^n_j + m * u^n_{j+1} \tag{6.12a}$$

where $\quad k = \dfrac{Cr}{2}\sqrt{\dfrac{g}{h_0}}, \quad m = -\dfrac{Cr}{2}\sqrt{\dfrac{g}{h_0}}, \quad l = 1 \quad$ and $\quad Cr = c\dfrac{\Delta t}{\Delta x}$

In a similar manner (6.11) can be reformulated as follows:

$$u^{n+1}_{j+1} = u^n_{j+1} - \frac{g\Delta t}{2\Delta x}(h'^{n+1}_{j+2} - h'^{n+1}_j) \tag{6.13}$$

Once again multiplying both the numerator and denominator of the second term in the right hand side of (6.13) by the still water depth h_0 and substituting the square of celerity, c^2, for (gh_0), (6.13) can be re-written in the following form:

$$u^{n+1}_{j+1} = p * h'^{n+1}_j + q * u^n_{j+1} + r * h'^{n+1}_{j+2} \tag{6.13a}$$

where $\quad p = \dfrac{Cr}{2}\sqrt{\dfrac{h_0}{g}}, \quad r = -\dfrac{Cr}{2}\sqrt{\dfrac{h_0}{g}}, \quad q = 1 \quad$ and $\quad Cr = c\dfrac{\Delta t}{\Delta x}$

In the difference schemes (6.12) and (6.13), the value of the dependent variables at one time is expressed as an explicit function of the value of the dependent variables at the earlier time. Such a scheme is then traditionally called an explicit difference scheme. More strictly speaking, it is an explicit scheme when it is non-recursive, i.e. it does not have to be solved in one particular order in space, so that it has no algorithmic structure (Abbott and Minns, 1998). In a similar manner, (6.5) and (6.7), which are the same linearised equations but described in terms of only one variable, h', can also be discretised as a leapfrog scheme and represented explicitly in a finite difference form, appearing, respectively, as follows:

$$h'^{n+1}_j = Cr^2 h'^n_{j+1} + 2(1 - Cr^2)h'^n_j + Cr^2 h'^n_{j-1} - h'^{n-1}_j \tag{6.14}$$

$$h'^{n+1}_{j,k} = Cr^2(h'^n_{j+1,k} + h'^n_{j-1,k} + h'^n_{j,k+1} + h'^n_{j,k-1}) + 2(1 - 2Cr^2)h'^n_{j,k} - h'^{n-1}_{j,k} \tag{6.15}$$

These are the forms of the explicit schemes that should be learnt from data in the present study.

6.3.2 The simple implicit formulation used as a test case

The nearly-horizontal flow Equations (6.1) and (6.2) can also be described in a so-called 'algorithmic' form as follows:

$$h\frac{\partial u}{\partial t} - u\frac{\partial h}{\partial t} - (u^2 - gh)\frac{\partial h}{\partial x} = 0 \tag{6.16}$$

$$g\frac{\partial h}{\partial t} - u\frac{\partial h}{\partial t} - (u^2 - gh)\frac{\partial u}{\partial x} = 0 \tag{6.17}$$

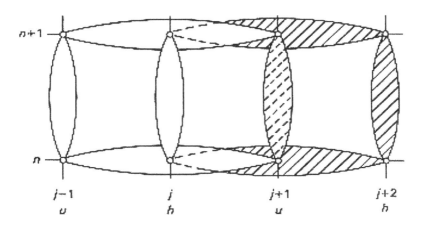

Figure 6.3. Schematisation of the Abbott Ionescu staggered scheme [adapted from Abbott and Basco, 1989].

Using centred space and time differences as in Figure 6.3. to discretise the terms in these equations and rearranging them leads to the Abbott Ionescu scheme (1967):

$$A1_j h_{j+1}^{n+1} + B1_j u_j^{n+1} + C1_j h_{j-1}^{n+1} = D1_j \tag{6.18}$$

$$A2_{j+1} u_{j+2}^{n+1} + B2_{j+1} h_{j+1}^{n+1} + C2_{j+1} u_j^{n+1} = D2_{j+1} \tag{6.19}$$

For the solution of (6.18) and (6.19) in sub-critical flows, one can introduce the relation:

$$h_{j+1}^{n+1} = E1_j u_j^{n+1} + F1_j \tag{6.20}$$

$$u_{j+2}^{n+1} = E2_{j+1} h_{j+1}^{n+1} + F2_{j+1} \tag{6.21}$$

with the corresponding recurrence relations:

$$E1_{j-1} = \frac{-C2_j}{A2_j E2_j + B2_j} \qquad F1_{j-1} = \frac{D2_j - A2_j F2_j}{A2_j E2_j + B2_j} \qquad (6.22)$$

$$E2_{j-1} = \frac{-C1_j}{A1_j E1_j + B1_j} \qquad F2_{j-1} = \frac{D1_j - A1_j F1_j}{A1_j E1_j + B1_j} \qquad (6.23)$$

These formulations are discussed in detail in Abbott and Basco (1989) and in Abbott and Minns (1998)

6.4   The Formulation of the Problem in the New Paradigm

As mentioned above, one essential component in the development of the newly envisaged modelling paradigm is that of allowing models to 'construct themselves', more or less automatically, by learning from existing numerical-hydraulic models and available (field) measurements. In the earlier chapters only the possibility of realising the learning process has been investigated, and even then this has been restricted to learning from data streams that have themselves been generated from existing numerical models for specific geographical locations while using the simplest tools for automated learning - artificial neural networks - that are currently available (Dibike et al., 1999b; Dibike and Abbott, 1999).   A next essential step has been to extend this work to encompass schemes that can be applied over arbitrary bathymetries with distances and time steps that are also arbitrary, consistent with the inherent possibilities of numerical stability of the representation. Since the purpose so far is only to establish the feasibility of such methods, the first step towards such generic systems that forms the subject of this study continues to be restricted to learning schemes that mimic existing and already well-established numerical schemes. The purpose has been to establish the possibilities and limits of the new paradigm by comparing its productions with the results of the earlier paradigm, as sketched above for the simplest possible cases. The present study is also concerned with establishing how far the artificial neural network technologies can continue to provide acceptable results.

Previous works (see, for example, Minns, 1998) demonstrated that, in the simple case of pure advection with constant velocity, a linear ANN is capable of learning the exact solution by actually restoring the advection equation from the weights of the ANNs. Promising results were also obtained in the case of short-period wave propagation as described in Chapter 3 (also see Dibike et al., 1999a). The present study, however, is directed towards investigating the possibility of using trained ANNs as part of computational schemes, corresponding to numerical operators, which could lead to the development of a generic hydraulic model using artificial neural networks from (in the present case, synthetic) data.

6.4.1 Experimental studies on the new paradigm with explicit formulation

First, a hypothetical rectangular channel of 10,000m length and 10m width with a horizontal bed was considered, as schematised in Figure 6.4. The channel was taken as closed at both sides and bed resistance was ignored. Flow data was then generated using a conventional numerical model. When the simulation started, the water surface was taken to be inclined, with a 9m water depth at one end and 11m at the other, with the intention of establishing a seiching motion along the channel. Different bottom configurations and grid spacings were considered as explained in the following sections:

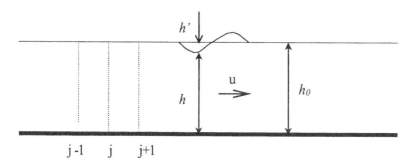

Figure 6.4. Schematic view of the one-dimensional channel considered.

Case 1: One-dimensional flow in a channel with a horizontal bottom and a uniform grid spacing

The first set of training experiments was performed for a uniform grid spacing, Δx, but with different time steps, Δt. However, the Courant number was always kept below unity because the input-output mapping was of explicit form. Since a linear problem was considered, and also in order to make the analysis easier, two-layered feed-forward neural networks with linear transfer functions were employed. When both flow variables, h and u, were considered, two networks had to be trained for each simulation, since there were two sets of operators in the numerical schemes (10) and (11). For the first network, $u(j-1)$, $h'(j)$

and $u(j+1)$ at time level n were taken as input while the $h'(j)$ at the $(n+1)th$ time level was taken as an output as shown on Figure 6.5a. For the second network, however, $u(j+1)$ at time level n and $h'(j)$ and $h'(j+2)$ at time level $n+1$ were taken as input, while the $u(j+1)$ at the $(n+1)th$ time level were taken as output as shown on Figure 6.5b. This meant that the first network had to solve h' values at alternate grid points along the channel and then the second network had to use these values to solve the u values at the intermediate grid points along the channel.

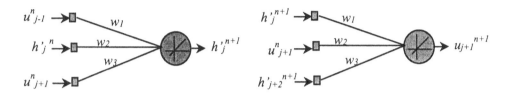

Figure 6.5 (a & b). ANN as numerical operators on a staggered grid.

After training the network with different combinations of parameter values, an acceptable performance was obtained in that the verification output data was reproduced with an acceptable accuracy. RMSEs of 0.000023 and 0.000072 were obtained for h' and u respectively. The weights of the network's connections obtained through the training corresponded to the coefficients that were multiplying the input variables in the original scheme and hence were directly comparable to the coefficients k, l, m, p, q and r of the finite different schemes described in (6.12a) and (6.13a). This training exercise was repeated for the different grid spacings (or Cr values) considered. Figure 6.6. shows a typical performance for different Courant numbers when the numerical coefficients are compared with the coefficients obtained from the weights of the ANNs.

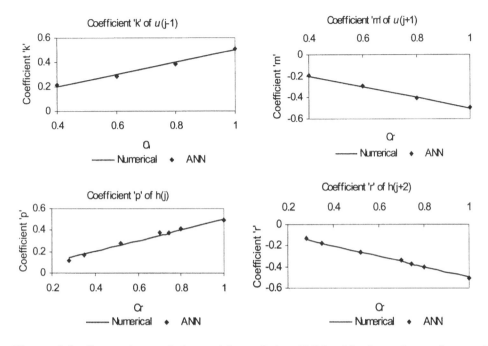

Figure 6.6. Comparison of the weights of the ANN with the values of numerical coefficients in (6.12a) and (6.13a).

In a similar experiment, a network was trained with only a single flow variable h' from the same simulation results as those used in the previous case. The variable h' at grid points $j-1, j,$ and $j+1$ at time level n and at grid point j at time level $n-1$ were considered as input to the network, while the h' at grid point j at time level $n+1$ was considered as output, in a way similar to that used in the case of the leap-frog operator (see Fig. 6.7).

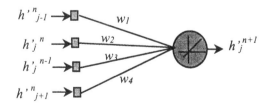

Figure 6.7. ANN as leap-frog operator.

Once again the training experiment was done for different grid spacings, while the Courant number was always kept below unity. The weights in the final configuration of the network's connections were compared with the coefficients of the finite different scheme

(6.14) corresponding to the different grid spacings (or Courant numbers) considered, and typical results are shown in Figure 6.8.

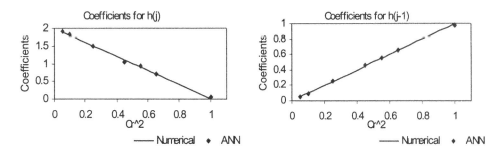

Figure 6.8. Comparison of weight of ANN with the value of numerical coefficient in (6.14).

In all the above cases, the plots of numerical coefficients obtained from the weights of the ANN fitted very well with those obtained from the finite difference schemes corresponding to the different Courant numbers. Nevertheless, in order to make the analysis more complete, the staggered grid numerical operators in the computer program were replaced by these trained networks and the simulations were reproduced. The water surface and velocity profiles obtained along the channel at a particular time level as well as the time series of water levels at a representative point were then plotted against the corresponding values obtained with the numerical model, as shown on Figures 6.9. to 6.12.

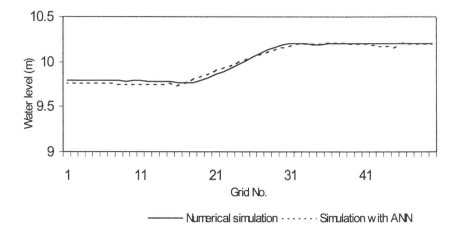

Figure 6.9. Comparison of Water level profiles in the channel after 4000 seconds.

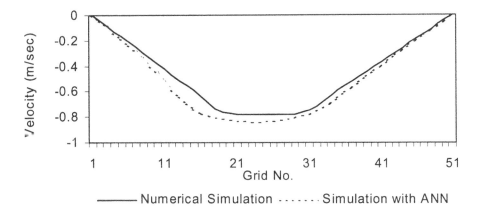

Figure 6.10. Comparison of the horizontal velocity distribution along the channel after 4000 seconds.

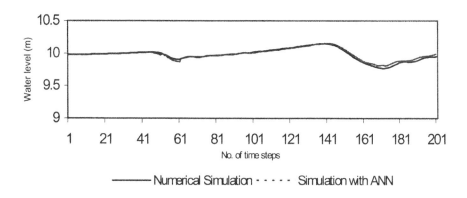

Figure 6.11. Comparison of time series of water level at the middle of the channel.

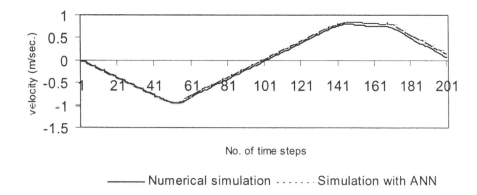

Figure 6.12. Comparison of time series of the flow velocity at the middle of the channel.

The water surface and velocity plots on Figures 6.9. and 6.10 show good agreement between the numerical simulations and simulations made by the ANNs. However, these plots do not by any means fit perfectly. In particular, there seems to be some loss of kinetic energy inherent in the results shown in Figure 6.10. Moreover the differences (or errors) in the time series of water level and velocity on Figures 6.11. and 6.12. continued to increase in time. Since the output of one network at one time level serves as an input to other networks at the next time level, the resulted increasing departure from the required values with time seems to be due to an accumulation of error.

However, these results can still be improved by a further training of the ANN employed, such as with additional hidden neurons and by using extended training data. The computation with ANNs also took a relatively longer time since the trained network had to run at every grid point. However, this should not be a problem when this approach is ultimately applied on non-structured grids, where the number of points in the flow field which are considered for simulation can be expected to be smaller in number.

It should also be noted that the ANNs took no account of the physical constraints, or laws, that are known to apply in such cases, such as are expressed as conservation laws of mass, momentum and (or) energy. The introduction of new methods of physically constrained data mining may make it possible to use constraints to further increase the accuracy of these methods substantially (see Babovic and Keijzer, 2000)

Case 2: One-dimensional flow in a channel with uneven bottom

The study has been extended to cover flow conditions in channels with non-horizontal beds and also to cover flow simulations with variable grid spacings. In the case where the channel bottom was not horizontal, the term h_0 and hence the wave celerity c along the channel were no more constant (see Fig. 6.13.).

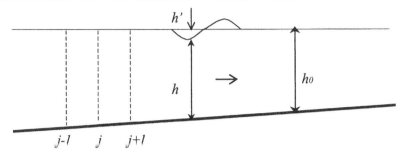

Figure 6.13. Schematic view of the one-dimensional channel with uneven bottom.

In order to train a linear network on the simulation result, (6.12) was rearranged into (6.24) and the data was prepared to feed the network accordingly.

$$\left(\frac{\Delta t}{\Delta x}\right)^2 h'^n_{j+1} - 2\left(\frac{\Delta t}{\Delta x}\right)^2 h'^n_j + \left(\frac{\Delta t}{\Delta x}\right)^2 h'^n_{j-1} = \frac{1}{gh_{0j}}\left(h'^{n+1}_j - 2h'^n_j + h'^{n-1}_j\right) \qquad (6.24)$$

This meant that the value of three h' terms at time level n were provided as input to the network, while the results of the expression at the right hand side of (6.24) were provided as output from the network. After the networks were trained with data obtained from model simulations with different $\Delta t/\Delta x$ ratios, the weights were compared with the corresponding numerical coefficients in (6.14). Typical results are shown in Figure 6.14.

Once the values of the expression at the right hand side of (6.24) were calculated by the network from the values of the variable h' at time level n, then its value at time level $n+1$ could be calculated at each grid point, by rearranging the terms on the right hand side of the same expression (6.24).

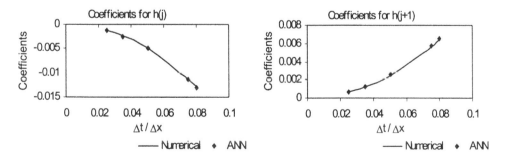

Figure 6.14. Plots of numerical-analysis derived coefficients with respect to Cr for channels with uneven bottom.

Case 3: one-dimensional flow in a channel with horizontal bottom and a variable grid spacing

If there is a change in the grid spacing along the channel, it is not possible to maintain a constant Cr. In order to incorporate this information into the linear network, the coefficients in (6.12) were rearranged and were rewritten as in (6.25):

$$\Delta t^2 h'^{n}_{j+1} - 2\Delta t^2 h'^{n}_{j} + \Delta t^2 h'^{n}_{j-1} = \frac{\Delta x_{j-1} \Delta x_j}{gh_{0j}} \left(h'^{n+1}_{j} - 2h'^{n}_{j} + h'^{n-1}_{j} \right) \qquad (6.25)$$

The input and output data were then prepared accordingly, corresponding to different Δt values, and the networks were then trained. Finally, the weights were compared with the corresponding numerical coefficients on the left hand side of (6.25). Some typical results are shown in Figure 6.15. Once again, the values of the expression at the right hand side of (6.25) were calculated by the network from the values of the variable h' at time n, so that their values at time level $n+1$ at each grid point could then be calculated by rearranging the terms on the right hand side of (6.25).

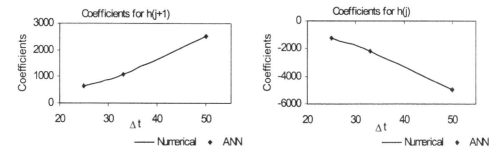

Figure 6.15. Plots of numerical-analysis derived coefficients with respect to Cr for channels with a variable grid spacing.

6.4.2   Experimental studies on the new paradigm with an implicit formulation

The next case considered was that of the implicit formulation, as introduced in § 6.3.2. Since implicit schemes are intrinsically recurrent, in this particular case the ANNs were required to model the recurrent relations in (6.22) and (6.23). In order to generate training data, one-dimensional hydrodynamic simulations based on Equations (6.16) and (6.17) were performed. However, not only the variables u and h, but also the coefficients in the intermediate steps, were written in a text file and prepared in such a way as to train the ANNs. As the recurrent relation is non linear, three-layer, feed-forward networks with tan-hyperbolic transfer functions in the hidden and output layers were employed. Two networks were trained for each simulation. For the first network, all the coefficients from *A1* to *F1* at grid point *j* were taken as inputs and the corresponding values of *E2* and *F2* at grid point *j-1* were taken as outputs. Similarly, the inputs to the second network were the coefficients from *A2* to *F2* at grid point *j*, while *E1* and *F1* at grid point *j-1* were the outputs. The networks were able to approximate the recurrence relations quite well, with RMSEs of 0.0007 and 0.0012 corresponding to (6.22) and (6.23) respectively.

In order to test the performances of these trained networks, they were once again imbedded in the program in such a way that, once the coefficients *A1* to *D1* and *A2* to *D2* had been calculated at alternating grid points, these two trained networks could then be run alternatively from one to the other boundary, thus functioning as numerical operators in place of (6.22) and (6.23). The simulation result obtained using these operators was compared with the one using a conventional numerical operator as shown in Figure 6.16.

Although Figures. 6.16. and 6.17. show a reasonable agreement between the simulation results obtained by the conventional method which uses the recursive relations in Equations (6.22) and (6.23), and the one which uses ANNs instead to calculate the intermediate coefficients, the difference between the two is more than the one obtained with the explicit formulation, as shown above in Figures 6.9. and 6.11. This could be attributed to an accumulation of error through the recursive process, since the output from one network at a given grid point is taken to be an input to the next network at the next grid point. The computation with ANNs also took a relatively longer time since the trained network had to run at every grid point. Nonetheless, this approach promises the additional advantage that it could be implemented with greater time steps (Cr > 1) because it is formulated in an implicit manner.

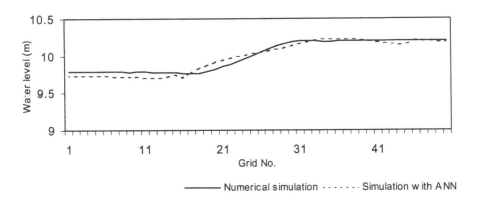

Figure 6.16. Comparison of Water level profiles in the channel after 4000 seconds.

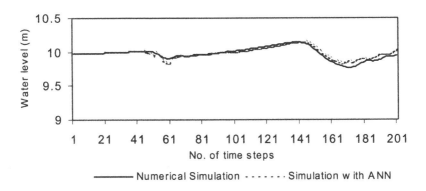

Figure 6.17. Comparison of time series of water level at the middle of the channel.

6.4.3   Experimental studies on two-dimensional flow in a rectangular basin

To extend this preliminary study to a two-dimensional flow case, a rectangular basin of 1500m by 2000m was considered. The basin was closed at the two longest sides, while it had partially open boundaries at the opposite corners of the other two sides, as shown in Figures 6.18., in such a way that a truly two-dimensional flow could be obtained. The basin had an initial depth of 10m and a wave of constant form with 0.5m amplitude and a wave period of 5 seconds was introduced at the upstream open boundary while the downstream boundary was kept at a constant depth of 10m. A grid spacing of 50 meters in both the x and y direction and a time step of 5 seconds was chosen to keep the Courant number close to unity.

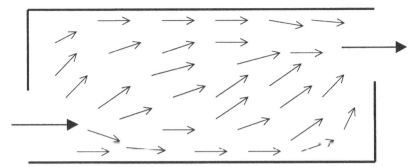

Figure 6.18. Plan of the rectangular basin for 2D flow simulation.

The flow in the basin was simulated using an alternating direction implicit (ADI) algorithm and the required flow data was generated. This flow data was then used to train a linear-transfer function ANN with $h'(j,k)$, $h'(j-1,k)$, $h'(j+1,k)$, $h'(j, k-1)$, $h'(j,k+1)$, all at time level n, and $h'(j, k)$ at time level $n-1$, as input to the network, and $h(j, k)$ at time level $n+1$ as an output (see Fig. 6.19) in a way similar to that employed in the case of the two-dimensional leap-frog scheme used to derive Equation (6.15).

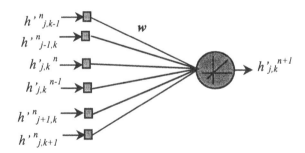

Figure 6.19. ANN as a two-dimensional leap-frog operator

The flow simulation and the corresponding training of ANN with the resulting flow data was repeated with successively decreasing time steps to keep Cr less than unity. The values of the weights of the ANNs and the coefficients of the corresponding finite difference scheme represented in Equation (6.15) were then plotted with respect to Cr for each input variable as shown in Figures 6.20. Once again the plots of numerical coefficients obtained from the weights of ANN fit very well with the ones obtained from the finite difference schemes corresponding to the different Courant numbers.

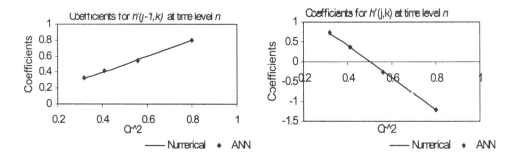

Figure 6.20. Plots of numerical coefficient with respect to Cr for simulations with variable grid spacing.

6.5 Discussion

The possibilities opened up by the new paradigm to allow models to be constructed by learning from existing numerical-hydraulic models, was further investigated by extending previous works to encompass schemes that can be applied over arbitrary bathymetries with distances and time steps that are also arbitrary. For the cases considered in this study, the artificial neural network provided encouraging results. This has been demonstrated for different cases of one and two dimensional flow problems by comparing the weights of the network's connections with the coefficients of the corresponding finite different schemes with different time steps (and hence different Courant numbers). The study also included cases with variable bed levels and schemes with variable grid spacing, and good results were obtained in these cases also.

Moreover, it was demonstrated that well-trained networks can be substituted in place of the finite difference schemes in the hydrodynamic model formulation and performed in the manner of numerical operators. In this case, models with trained ANNs as explicit numerical operators perform better than the ones with ANNs as implicit operators. However, the implicit operators have the extra advantage that they can be implemented with relatively longer time steps (and hence $Cr > 1$). In general, flow simulations with trained ANNs as numerical operators required longer running times. However, this approach is still very useful when implemented for model simulations using non-structured grids, where the number of points of interest, and hence points to be considered for the simulations, usually becomes smaller and a reasonable amount of historical (or synthetic) data are available at those locations. In such cases, ANNs can be trained based on the data at selected points in the model area and the corresponding boundary conditions. These trained networks can then be used to emulate numerical operators for subsequent model simulations in the same area.

Chapter 7

Summary and Conclusions

7.1 Summary

Conventional modelling approaches in hydraulics use mathematical relationships which are derived from a few fundamental principles, such as the conservation of mass, momentum and energy. This usually leads to the posing of hydraulic problems in terms of the solution of systems of non-linear partial differential equations. If the initial and boundary conditions of a particular hydraulic system, along with its geometry, are provided as inputs, such equations can be solved numerically to provide, as an output, the variables which describe the state of the system at subsequent time levels. Once such a model is built, it can then be used to simulate the behaviour of the particular system which it represents. However, even if the mathematical representations used to develop such a model adequately describe the physical processes in the system, the numerical values of the various parameters associated with such representations have to be determined by calibrating the model against measured data.

Recent advances in research in the area of model identification have revealed other, alternative, approaches for inducing models from data. These approaches are usually based on learning systems which can determine the relationships between the input and output variables from historical data presented to them, without resorting to describing these relationships explicitly in mathematical form. One such learning system, which has been the subject of most of this thesis, is that of artificial neural networks. A neural network is a system of processing units or neurons with simple multiplicative, additive and other functional elements that are connected into a network through a set of *weights* such that this network provides a relationship between the input and the output data sets. The mode of functioning of the network is determined by the network's architecture, the magnitudes of the weights and the mode of operation of the processing elements.

Artificial neural networks, and in particular the multilayer perceptron and recurrent networks, have gained an immense popularity in recent years as data-driven modelling tools in all manner of areas of application. They are now widely used instruments in hydraulics and hydrology since they perform well in most cases of practical interest. However, there is a large number of other learning systems which can also be used to induce models of physical phenomena from available data, some biologically motivated (such as most evolutionary algorithms) and others derived from statistical learning theories, such as support vector machines (SVMs). Although the use of SV methods in applications has begun only recently, a considerable number of researchers have already reported a state of the art performance in a variety of applications, such as pattern recognition, regression estimation, and time series prediction. An interesting property of the SVM is that it is an approximate implementation of the Structural Risk Minimisation (SRM) induction principle that aims at minimising a bound on the generalisation error of a model and has been shown to be superior to the more traditional Empirical Risk Minimisation (ERM) principle employed by many other modelling techniques. The applicability of SVMs for higher dimensional regression estimation is demonstrated in Chapter 3 by applying the method for two modelling (regression approximation) problems, one in hydrology and the other in hydraulic engineering. The SVM approach to rainfall-runoff modelling has been shown to be a better alternative to the traditional conceptual model to which the results were compared. The SVMs were found to generalise better by giving a more accurate prediction of runoff on test data. SVMs were also shown to provide a good performance for the prediction of forces on vertical hydraulic structures due to dynamic waves.

Despite the powerful processing capabilities of neural network systems, interpretation of their internal knowledge representation is largely incomprehensible to humans. Neural networks do not themselves provide an explanation of their mode of functioning as part of their information processing capability; the knowledge that they have gained through training is only stored implicitly in their weights. However, it has been argued in Chapter 4 that the knowledge imbedded in a (generalised) neural network system can indeed be represented in a symbolic form, such as in the form of a system of partial differential equations and that ANNs can encapsulate the same knowledge or exhibit the same semantic content as do continuum equations. This has been demonstrated by analysing the weights of the ANNs that were trained with a synthetic data set generated by the Boussinesq wave module of a hydrodynamic modelling system (MIKE 21 BW) and by then actually recovering the continuum Boussinesq-type partial differential equations from this data set. This attempt to derive numerical schemes and thus partial differential equations directly from data using ANNs may serve to induce some confidence in the predictive abilities and thus the applicability of ANNs as the bases of reliable data-driven modelling procedures. Beyond this, however, it has also been argued that if numerical solutions of specific partial differential equations are used to provide the data sets that are to train the ANNs, and the resulting trained weights of the ANNs are able to reinstate the original differential equations, then data taken from nature (that may be treated to minimise noise and other distorting influences) should just as well provide ANNs that can in turn be used to produce partial differential equations. This form of knowledge representation, as partial differential equations, is however already generally accepted as a means for encapsulating our knowledge of the behaviour of specific processes in hydraulics. Thus ANN weighting can serve just as well for knowledge discovery and knowledge encapsulation purposes.

The results of the investigation in Chapter 5 reinforced the above argument by showing that agents constructed exclusively from ANNs could be used as non linear dynamic system models to encapsulate site-specific knowledge and data and to emulate the temporal sequence of states observed in a two dimensional model of a specific geographic area, namely Donegal Bay in Ireland. Even though this model was constructed using a set of different input series at the open boundaries, the study was site-specific to a region with a relatively simple geometry. It is to be anticipated that increasing the generality of the problem by extending it to provide general, non-site-specific, modelling arrangements suited to more complicated bathymetries will necessitate the use of agents that are considerably more heterogeneous in their composition.

In Chapter 6, the possibilities opened up by the ANN paradigm that allows models to be built by learning from data generated by using existing numerical-hydraulic models, was investigated by extending previous work to encompass schemes that can be applied over arbitrary bathymetries with distances and time steps that are also arbitrary. For the relatively simple cases considered in this study, artificial neural networks continue to provide good results. The study also included cases with variable bed levels and schemes with variable grid spacing: also here, promising results were obtained. Moreover, it was demonstrated that the well-trained networks could be substituted in place of the conventional finite difference schemes in the hydrodynamic model formulation and could perform very much like numerical operators. In general, flow simulations with trained ANNs as numerical operators required relatively longer computer running times. However, this approach is still very useful when implemented for model simulations using non-structured grids, where the number of points of interest, and hence the number of points to be considered for a simulation, may be relatively small, and a reasonable amount of historical (or synthetic) data are available at these locations. In such cases, it has been shown that ANNs can be trained based on the data at selected points in the model area and the corresponding boundary conditions and these trained networks can then be used as numerical operators for subsequent model simulations in the same area. To proceed beyond this, to the generation of learning systems that will provide general non-structured grid models for arbitrary bathymetries, appears to be quite feasible. The effort required to tackle any such much-extended learning process could be realised using the generalisation towards the SVM approach, tailoring the computational kernel to the specific problem at hand.

7.2   Conclusion

In general, *model induction from data* aims at providing tools to facilitate the conversion of data into other forms that provide better ways of simulating the physical processes that generated or produced these data. When combined with the already available understanding of the physical process, these new models, result in an improved formulation of modelling problems and so provide an improved predictive capability (Abbott et al., 2001). The work presented in this thesis, therefore, demonstrates this approach as one possible direction of progress towards the next generation of computational hydraulic models.

Artificial neural networks are very simple mathematical operators that can be implemented with relative ease and run very fast in most practical applications. Moreover, the ANN's ability to 'learn' from data makes it very attractive, especially when the characteristics of the physical processes to be modelled (or some part of these processes) are not yet well established in physical terms. They can be implemented as stand-alone systems, as systems that can model relationships between the different variables in a hydraulic system, or as part of a more complex system representing a particular component of a more general hydroinformatics system. For instance, recent studies (Babovic et al., 2000) have demonstrated that the forecasting ability of numerical models can be significantly improved by using such hybrid systems whereby the data-driven models such as ANNs and SVMs can be used as data assimilation tools for updating the output variables (or for predicting the errors). By using these techniques one can assimilate the forecast of the numerical model at points of interest with the latest observed data in order to predict the error and obtain an improved forecast. In another study, Boogaard et al. (2000b) has also demonstrated how an auto-regressive neural network (ARNN) can be used for modelling dynamic processes. Moreover, in order to account for uncertainties in the model's state, external forcings and the model parameters, the trained deterministic ARNN-model is embeded into a stochastic environment. Similar to operational conceptual numerical models, this extended data driven model can then be combined with on-line data assimilation facilities. In this way the ARNN is capable of dealing with changing conditions and it's skill for accurate short to medium term predictions is significantly improved so that these models are quite well suited for application under real time operational conditions.

Several neural network simulation packages are available nowadays, that translate ANN solutions into a standard computer code. These computer codes (ANN agents) can then be employed as embedded functions using standard calls within existing numerical models, or coupled externally to the models to form an integrated part of a larger modelling system. If these ANN agents are equipped with an on-line learning capability (i.e. the ability to learn from on-line data that they receive from continued measurements of environmental variables), then the resulting modelling system will exhibit elements of intelligent behaviour during its operation, in a way contributing towards realising a new generation of computational engines in hydraulics and hydrology.

Model inductie uit data: naar de volgende generatie rekenmethodieken in hydraulica en hydrologie

Samenvatting

Conventionele modelbenaderingen in de hydraulica en de hydrologie gebruiken wiskundige relaties die afgeleid zijn van een aantal basisprincipes zoals behoud van massa, impuls en energie. In de praktijk leidt dit tot het formuleren van hydraulische problemen in termen van de oplossing van systemen van niet-lineaire partiële differentiaal vergelijkingen. Als de begin- en randvoorwaarden van een specifiek hydraulisch probleem, inclusief de geometrie, als input worden ingevoerd kunnen dergelijke vergelijkingen numeriek worden opgelost met als output de variabelen die de toestand van het system in opeenvolgende tijdstippen beschrijven. Wanneer een dergelijk model is gebouwd, kan het worden gebruikt om het gedrag van een specifiek probleem te simuleren. Echter, zelfs wanneer de mathematische representaties die gebruikt worden voor de ontwikkeling van zo'n model de fysische processen in het systeem adequaat beschrijven, moeten de numerieke waarden van de verschillende parameters, die gekoppeld zijn aan die representaties, vaak vastgesteld worden door middel van calibratie van het numerieke model op basis van gemeten data.

Recente ontwikkelingen op het gebied van onderzoek naar modelidentificatie hebben andere, alternatieve, benaderingen opgeleverd voor het ontwikkelen van modellen uit data. Deze benaderingen zijn meestal gebaseerd op leersystemen die de relaties tussen input en output variabelen kunnen vaststellen uit historische gegevens, zonder de noodzaak van het expliciet beschrijven van deze relaties in een mathematische vorm. Een van de leersystemen, die centraal staat in het grootste deel van deze thesis, is die van 'artificial neural networks' (ANNs). Een neural network is een systeem van proceselementen of wel 'neurons' met eenvoudige vermenigvuldigings-, sommerings- en andere functionele eigenschappen die verbonden zijn in een netwerk door middel van een set *gewichten* zodanig dat dit netwerk een relatie beschrijft tussen de input en ouptut data sets. De wijze van functioneren van het netwerk wordt bepaald door de architectuur van het netwerk, de zwaarte van de gewichten en de wijze van beheer van de proceselementen.

ANNs, en in het bijzonder de meerlagen formulering en recurrente netwerken hebben de afgelopen jaren enorm aan populariteit gewonnen als datagestuurde modellen in alle mogelijke vormen van toepassing. Ze zijn nu wijd verpreide instrumenten in de waterbouwkunde en hydrologie omdat ze goed functioneren in de meeste praktijk situtaties. Echter, er is een groot aantal andere leersystemen die ook gebruikt kunnen worden om modellen te genereren vanuit fysische verschijnselen op basis van verschillende data, sommige biologisch (zoals de meeste evolutie algorithmen) en andere weer van statistische leertheorieën, zoals 'support vector machines' (SVMs). Hoewel de toepassing van SV methoden zich slechts in het beginstadium bevindt heeft een groot aantal onderzoekers inmiddels gerapporteerd over resultaten van verschillende toepassingen zoals patroonherkenning, regressie schattingen en tijdreeks voorspellingen. Een interessante eigenschap van de SVM is dat het een benaderende uitvoering is van het Structural Risk Minimisation (SRM) inductie principe, dat gericht is op het minimaliseren van de limiet van de generalisatiefout van een model en heeft aangetoond beter te zijn dan het meer traditionele Empirical Risk Minimisation (ERM) principe, dat door vele andere model technieken wordt gebruikt. De toepasbaarheid van SVMs voor hogere dimensionale regressie schatting wordt aangetoond in Hoofdstuk 3 door het toepassen van de methode voor twee (regressie benaderings)modelproblemen: een in hydrologie en een in de waterbouwkunde. De SVM benadering voor neerslag-afvoer-modellering blijkt een beter alternatief te zijn dan het traditionele conceptuele model waarmee de resultaten worden vergeleken. De SVMs blijken beter te generaliseren door middel van meer accurate voorspellingen van afvoer op basis van testgegevens. SVMs blijken ook goede resultaten te geven voor de voorspelling van krachten op verticale waterbouwkundige constructies ten gevolge van dynamische golfbewegingen.

Ondanks de krachtige proceseigenschappen van ANNs, is de interpretatie van de interne kennisrepresentatie grotendeels onbegrijpelijk voor de mens. Neurale netwerken geven zelf geen verklaring voor de wijze van functioneren als deel van hun informatie procesvaardigheden; de kennis die is vergaard door middel van training is slechts impliciet opgeslagen in hun gewichten. Echter, in Hoofdstuk 4 wordt aangetoond dat, kennis die is ingebed in een veralgemeniseerd neuraal netwerk inderdaad ook op een symbolische wijze kan worden uitgedrukt, zoals in de vorm van een systeem van partiële differentiaal vergelijkingen en dat ANNs dezelfde kennis kunnen omvatten of dezelfde semantische inhoud vertonen als de continue (partiële differentiaal) vergelijkingen. Dit wordt aangetoond door de gewichten van de ANNs te analyseren die getraind zijn met een kunstmatige data set gegenereerd door de Boussinesq golf module van een hydrodynamisch modellerings-systeem (MIKE 21 BW), en door daarna dezelfde de continue Boussinesq-type partiële differentiaal vergelijkingen uit deze data sets te reconstrueren. Deze poging om numerieke schema's en dus partiële differentiaal vergelijkingen direct te verkrijgen uit data met behulp van ANNs, kan dienen om enig vertrouwen teweeg te brengen in het voorspellende vermogen en daarmee de toepassing van ANNs als de basis voor betrouwbare datagestuurde modelleringsprocedures. Behalve dit wordt ook beargumenteerd dat wanneer numerieke oplossingen van specifieke partiële differentiaal vergelijkingen gebruikt worden als data sets om ANNs te trainen, zodanig dat uit de getrainde gewichten van de ANNs de originele differentiaal vergelijkingen kunnen worden geconstrueerd, dan zou data die in de natuur is waargenomen (al dan niet gecorrigeerd om ruis en andere verstorende invloeden te minimaliseren) net zo goed ANNs kunnen leveren die op zich weer gebruikt kunnen worden om partiële differentiaal vergelijkingen te produceren. Kennisrepresentatie in de vorm van partiële differentiaal vergelijkingen is in het algemeen meer geaccepteerd als manier om onze kennis van het gedrag van specifieke processen in waterbouw in te kapselen. Maar op deze manier kunnen ANNs ook voor kennisontsluiting en kennisvastlegging dienen.

De resultaten van het onderzoek in Hoofdstuk 5 versterken bovenstaand argument door te tonen dat 'agents' die exclusief uit ANNs samengesteld zijn gebruikt konden worden als niet-lineair dynamische modelsysteem om locatie specifieke kennis en data vast te leggen en de tijdafhankelijke waargenomen stadia in een twee-dimensionaal model van een specifiek geografisch gebied, namelijk Donegal Bay in Ierland, te emuleren. Hoewel dit model vervaardigd was met gebruik van een reeks verschillende input condities op de open randen, was het onderzoek locatie gebonden aan een regio met een relatieve eenvoudige geometrie. Het is te verwachten dat een verdere veralgemenisering van deze aanpak mogelijk is door het uitbreiden naar algemene, niet locatie gebonden, modelleringsvormen en door het gebruik van agents voor meer gecompliceerde bathymetrieën en het gebruik van agents, die aanzienlijk meer heterogeen van samenstelling zijn.

In Hoofdstuk 6, worden de mogelijkheden van het ANN paradigma onderzocht om modellen te bouwen op basis van data gegenereerd met behulp van bestaande numerieke hydraulische modellen. Dit is gedaan door uitbreiding naar schema's die kunnen worden toegepast op arbitraire bathymetrieën met eveneens arbitraire afstanden en tijdstappen. Voor de relatief eenvoudige gevallen die in deze studie zijn beschouwd, blijft ANN goede resultaten geven. Het onderzoek omvatte ook voorbeelden met variabele bodem niveau's en schema's met variabele grid afstanden. Ook hier werden veelbelovende resultaten geboekt. Bovendien werd aangetoond dat de goedgetrainde netwerken konden worden vervangen door conventionele eindige differentie schema's in de hydrodynamische modelformulering en goed konden functioneren als numerieke operators. In het algemeen vereisen stromingssimulaties met getrainde ANNs als numerieke operatoren relatief lange rekentijden. Echter, deze benadering is nog steeds bijzonder goed bruikbaar indien toegepast op modelsituaties die gebruik maken van niet-gestructureerde rekenroosters waar het aantal meetpunten en daarmee het aantal punten dat in beschouwing moet worden genomen voor een simulatie relatief klein is, en wanneer een redelijk aantal historische gegevens beschikbaar is voor deze locaties. Voor zulke gevallen wordt aangetoond dat ANNs getrained kunnen worden op basis van data op geselecteerde punten in het modelgebied en de bijbehorende randvoorwaarden, en dat deze getrainde netwerken gebruikt kunnen worden als numerieke operatoren voor volgende model simulaties in hetzelfde gebied. Het lijkt zeer wel mogelijk om dit leerproces uit te breiden naar een generatie van systemen die algemene niet gestructureerde grid modellen voor arbitraire bathymetrieën zal leveren. De inspanning die nodig is om zo'n zeer uitgebreid leerproces op te zetten zou gerealiseerd kunnen worden door de generalisatie van de SVM benadering te gebruiken en de 'reken kern' af te stemmen op het specifieke probleem.

In het algemeen probeert *model inductie uit data* hulp te bieden aan de conversie van data naar andere vormen van representatie die op een betere manier de simulatie beschrijven van fysische processen waaruit de data zijn gegenereerd of geproduceerd. Gecombineerd met het al aanwezige begrip van de fysische processen kunnen deze nieuwe modellen resulteren in een verbeterde formulering van modelleringsproblemen en zullen zo mogelijk een beter voorspellend vermogen bieden (Abbott et al., 2001). Het werk in deze thesis laat deze benadering zien als een mogelijke richting voor de volgende generatie van numerieke hydraulische modellen.

Artificial neural networks zijn hele simpele wiskundige operatoren die vrij gemakkelijk kunnen worden toegepast en snel werken op de meeste praktische toepassingen. Bovendien is het vermogen van ANNs om te 'leren' van data erg aantrekkelijk, vooral wanneer de karakteristieken van de fysisiche processen die gemodeleerd moeten worden (of een deel van deze processen) in fysische termen nog niet goed vastgesteld zijn. Ze kunnen worden toegepast als alleenstaand systeem, als systeem dat vorm geeft aan verhoudingen tussen de verschillende variabelen in een hydraulisch systeem, of als deel van een meer complex system dat een bepaalde component van een meer algemeen hydroinformatisch systeem uitmaakt. Tegenwoordig zijn verscheidene neurale netwerk simulatiepakketten beschikbaar die ANN oplossingen vertalen naar een standaard computer code. Deze computer codes (ANN agents) kunnen dan worden opgenomen als ingesloten (standaard) functies binnen bestaande numerieke modellen, dan wel extern gekoppeld worden aan deze modellen om een geïntegreerd geheel te vormen met een groter modelleringssysteem. Als deze ANN agents uitgerust zijn met een on-line leervermogen (bijvoorbeeld het vermogen om te leren van on-line data die zij ontvangen van lopende metingen van de omgevingsvariabelen), dan zal het modelleringssysteem als resultaat elementen van intelligent gedrag vertonen tijdens het functioneren, en zo bijdragen aan het realiseren van een nieuwe generatie van numerieke rekenharten in waterbouwkunde en hydrologie.

Curriculum Vitae

Yonas Berhan Dibike was born on February 20, 1969 in Addis Ababa, Ethiopia. In 1991 he received his B.Sc. degree in Hydraulic Engineering with distinction from Arbaminch Water Technology Institute (AWTI), Arbaminch, Ethiopia. From 1992 till 1994 he worked first as graduate assistance and then as an assistant lecturer at the same institute. In 1995 he spent eleven months attending a post graduate course in Water Resources Engineering at Dar es Salaam University (UDSM), Dar es Salaam, Tanzania. In October 1995 he joined the postgraduate course in Hydraulic Engineering at the International Institute for Infrastructural, Hydraulic and Environmental Engineering (IHE), Delft, the Netherlands and in April 1997 he obtained his MSc degree in Hydroinformatics, with distinction. Since May 1997, he continued to work at IHE as a teaching assistant (AIO) and PhD fellow in hydroinformatics.

References

Abbott, M.B. (1966) *An Introduction to the Method of Characteristics*, Thames and Hudson, London.

Abbott, M.B., and Ionescu F. (1967) On the numerical computation of nearly horizontal flows, *Journal of Hydraulic research*, Vol. 5, pp. 97-117.

Abbott, M.B., Petersen, H.M., and Skovaard, (1978) On the numerical modelling of short waves in shallow water, *Journal of Hydraulic Research*, Vol. 16, pp. 173-203.

Abbott, M.B. (1979 // 1985) *Computational Hydraulics: Elements of the theory of free surface flows,* Pitman Publishing limited, London.

Abbott, M.B., and Basco, D.R. (1989) *Computational Fluid Dynamics*, Longman, London.

Abbott, M.B. (1991) *Hydroinformatics: informational technology and the aquatic environment*, Ashgate, Aldershot, and Brookfield, USA.

Abbott, M.B. (1993) The electronic encapsulation of knowledge in hydraulics, hydrology and water resources, *In Advances in Water Resources,* Vol. 16, pp. 21-39.

Abbott, M.B. (1997a) Engine 2000: Research in to the next generation of computational hydraulic modelling, *Proc. of the 27th Congress of the International Association for Hydraulic Research,* Vol. 2, pp. 859-864.

Abbott, M.B. (1997b) Range of Tidal Flow Modelling, *Journal of Hydraulic Engineering,* Vol. 123, pp. 257-276.

Abbott, M.B., and Dibike, Y.B. (1998) The symbolic representation of hydroinformatic processes using elements of category theory, *Hydroinformatics'98*, Babovic and Larsen (eds), Balkema, Rotterda, pp.1185-1192.

Abbott, M.B., and Minns, A.W. (1998) *Computational hydraulics,* 2nd edition, Ashgate, Aldershot, UK, and Brookfield, USA.

Abbott, M.B., Babovic V.M., and Cunge, J.A. (2001) Towards the hydraulics of the hydroinformatics era, *Journal of Hydraulic research,* Vol. 39, No. 4, pp. 339-349.

Abrahart, R.J. (1998) Neural Networks and the problem of accumulated error: embedded solution that offers new opportunities for modelling and testing, on *Proc. 3rd Int. Conf. In Hydroinformatics* , Babovic *et al* (eds), Vol. 2, pp. 725-731.

Andrews, R., Diederich J., and Tickle A. (1995) A Survey and Critique of Techniques For Extracting Rules From Trained Artificial Neural Networks, *Knowledge Based Systems,* Vol. 8, pp. 373-389.

ASCE Task Committee on Application of Artificial Neural Networks in Hydrology (2000a) Artificial Neural Networks in Hydrology. I: Preliminary Concepts, *ASCE Journal of Hydrologic Engineering*, Vol. 5, No. 2, pp. 115-123.

ASCE Task Committee on Application of Artificial Neural Networks in Hydrology (2000b) Artificial Neural Networks in Hydrology. II: Hydrologic Applications, *ASCE Journal of Hydrologic Engineering*, Vol. 5, No. 2, pp. 124-137.

Babovic, V., Keijzer, M., and Bundzel, M. (2000) From Global to Local Modelling: A Case Study in Error correction of Deterministic Models. *Hydroinformatics'2000*, Iowa, USA.

Babovic, V., and Keijzer, M. (2000) Genetic programming as a model induction engine, *Journal of Hydroinformatics,* Vol. 2, No. 1, pp. 35 – 60.

Babovic, V. (1996) *Emergence, Evolution, Intelligence; Hydroinformatics*, Ph.D Thesis, Balkema, Roterdam.

Babovic, V., and Abbott, M.B. (1997) The evolution of equation from hydraulic data (Parts I & II), *Journal of hydraulic research*, Vol. 35, No. 3, pp.397-430.

Babovic, V. (1997) An application of Artificial Neural Networks in Computational Hydraulics, *Proc. of the 27th Congress of the International Association for Hydraulic Research,* Vol. 2, pp. 865-870.

Boogaard, H.F.P., Gautam, D.K., and Mynett, A.E. (1998) Auto-Regressive Neural Networks for the modelling of time series, on *Proc. 3rd Int. Conf. In Hydroinformatics* , Babovic and Larsen (eds), Vol. 2, pp.

Boogaard, H.F.P., Mynett, A.E., and Heskes T. (2000a) Resampling Techniques for the assessment of uncertainities in parameters and predictions of calibrated models, *In Proc. of the 3rd International Conference on Hydroinformatics,* Iowa City, USA.

Boogaard, H.F.P., Mynett, A.E., and Heskes T. (2000b) On-line data assimilation in auto-regressive neural networks, *In Proc. of the 3rd International Conference on Hydroinformatics,* Iowa City, USA

Box, G.E.P., and Jenkins, G.M. (1976), *Time Series Analysis: Forecasting and Control,* Holden-Day, San Francisco.

Breitscheidel, A., Cser, J. and v.d Veer, P, (1998), Applicability of neural networks in hybrid water models, on *Proc. 3rd Int. Conf. In Hydroinformatics* , Babovic and Larsen (eds), Vol. 2, pp. 749-752.

Burges C.J.C. (1998) A Tutorial on Support Vector Machines for Pattern Recognition. In *Knowledge Discovery and Data Mining*, Vol. 2, No. 2, pp.1-43.

Burges C.J.C. (1999) Geometry and invariance in kernel based methods. In Schölkopf et al (eds.), *Advances in Kernel Methods : Support Vector Learning*, pp. 89-116, Cambridge, MA, MIT Press.

Cherkassky, V., and   Mulier, F. (1998) *Learning from Data: Concept, Theory and Methods.* John Wiley and Sons, New York.

Cloete, I. (2000), Knowledge-Based Neurocomputing: Past, Present and Future, In Cloete and Zurada (eds.), *Knowledge-Based Neurocomputing*, MIT Press.

Cortes, C., and Vapnik, V. (1995) Support vector networks, *Machine Learning*, No. 20, pp. 273-297.

Cristianini, N., and Shawe-Taylor, J. (2000) *An introduction to Support Vector Machines and other kernel-based methods*, Cambridge university press, UK.

Cunge, J.A., Holly, F.M., and Verwey, A. (1980*) Practical Aspects of computational River Hydraulics*, Pitman Publishing Inc., London.

Dibike Y.B., Velickov S., Solomatine D. and Abbott M.B., 2001, Model Induction with Support Vector Machines: Introduction and Applications, ASCE *Journal of Computing in Civil Engineering*, Vol. 15, No. 3, pp. 208- 216.

Dibike, Y.B., and Solomatine, D. (2001) River Flow Forecasting Using Artificial Neural Networks, *Journal of Physics and Chemistry of the Earth, Part B: Hydrology, Oceans and Atmosphere*, Vol. 26, No.1, pp. 1-8.

Dibike, Y.B. (2000) Machine Learning Paradigms for Rainfall-Runoff Modelling, *In Proc. of the 3rd International Conference on Hydroinformatics,* Iowa City, USA.

Dibike, Y.B., Velickov, S., and Solomatine, D. (2000) Support Vector Machines: review and applications in civil engineering, In Schleider and Zijderveld (eds.), *Artificial Intelligence Methods in Civil Enginnering Applications*, pp. 45-58.

Dibike, Y.B., and Abbott, M.B. (1999) Application of artificial neural networks to the simulation of two-dimensional flow, *Journal of Hydraulic research,* Vol. 37, No. 4, pp. 435-446.

Dibike, Y.B., Minns, A.W., and Abbott, M.B. (1999a), Application of Artificial Neural Networks to the Generation of wave Equations from Hydraulic Data., *Journal of Hydraulic research,* Vol. 37, No. 1, pp. 81-97.

Dibike, Y.B., Solomatine, D., and Abbott, M.B. (1999b). On the encapsulation of numerical-hydraulic models in artificial neural network, *Journal of Hydraulic research,* Vol. 37, No. 2, pp. 147-161.

Dingemans, M.W. (1997) *Water wave propagation over uneven bottoms* (part 1 & 2), World Scientific publishing, Singapore.

Drossu, R., and Obradovic, Z. (2000) Data mining Techniques for Designing Neural Network Time Series Predictors, In Cloete and Zurada (eds.), *Knowledge-Based Neurocomputing*, MIT Press.

Elman, J. (1990) Finding structure in time, *Cognitive Science*, No. 14, pp. 179-211.

Gautam, D.K., and Holz, K.-P. (2000) Neural network based system identification approach for the modelling of water resources and environmental systems, In Schleider and Zijderveld (eds.) *Artificial Intelligence Methods in Civil Enginnering Applications*, pp. 87-100.

Gent, M.R.A., and Boogaard, H.v.d. (1998) *Neural Network and Numerical Modelling of Forces on Vertical Structures*, MAST-PROVERBS report, Delft Hydraulics.

Goda, Y. (1985) *Random Seas and Design of Maritime Structures.* University of Tokyo Press.

Grabec, I. (1990) Emperical modelling of natural phenomena by a self-organizing system, *Proc. Neural Network Conference - 90*, Vol. 2, 529-532,

Gunn, S. (1998) Support Vector Machines for Classification and Regression. *ISIS Technical Report.*

Hall, M.J., and Minns, A.W. (1993) Rainfall-runoff modelling as a problem in artificial intelligence: experience with a neural network. In: *Proc. BHS 4th National Hydrology Symposium*, Cardiff, pp. 5.51-5.57.

Hall, M.J., Minns, A.W., and Ashrafuzzaman, A.K.M. (2000) Regional flood frequency analysis using artificial neural networks: a case study, *In Proc. of the 3rd International Conference on Hydroinformatics*, Iowa City, USA.

Hassoun, M.H. (1995) *Fundamentals of Artificial Neural Networks*. MIT.

Holland, J.H., Holyoak, K.J., Nisbett, R.E., and Thagard, P.R. (1986) *Induction: Processes of Inference, Learning and Discovery*, MIT press, Cambridge, Massachusetts.

Hopfield, J.J. (1982) Neural networks and physical systems with emergent collective computational abilities, *Proc. National Acadamy of Sciences USA*, No. 79, pp. 2554-2558.

Hornik, K., Stinchcombe, M., and White, H. (1989) Multilayer feedforward networks are universal approximators, *Neural Networks*, 2: 359-366.

Jordan, M.L. (1986) Attractor dynamics and parallelism in a connectionist sequential machine. *Proc. of the 8th conference of the Cognitive Science Society*, pp. 532-546.

Kachroo, R.K., Liang, G.C., and O'Connor, K.M. (1995) Intercomparison Study of Mathematical Models for River Flow Forecasting. *Report on Flood Forecasting Workshop*, Galway, Ireland.

Kachroo, R.K. (1992) River flow forecasting. Part 1: A discussion of the principles, In: *Journal of Hydrology*, 133, pp. 1-15.

Kohonen, T. (1982). Self-organized formation of topologically correct feature maps. Biological Cybernetics, No. 43, pp. 59 - 69.

Kohonen, T. (1995) *Self-Organizing Maps*. Springer Series in Information Sciences, Vol.80, Springer-Verlag.

Koza, J.R. (1992) *Genetic Programing: on the programming of computers by natural selection*, MIT, Cambridge, MA.

Lobbrecht, A.H., and Solomatine, D.P. (1999) Control of water levels in polder areas using neural networks and fuzzy adaptive systems. In: Water Industry Systems: Modelling and Optimization Applications, D.Savic, G. Walters (eds.), Vol.1, pp. 509-518.

Lorrai, M., and Sechi, G.M. (1995) Neural Nets for Modelling Rainfall-Runoff Transformations, In: *Water resources management*, Vol. 9, pp. 299-313.

Madala, H.R., and Ivakhnenko, A.G. (1994) *Inductive learning algorithms for complex systems modeling*, CRC Press, U.S.A.

Madsen, P.A., Murray, R., and Sorensen, O.R. (1991) A new form of Boussinesq equations with improved linear dispertion characteristics. *Coastal Engineering*, pp. 371-388.

Malsburg, C. von der (1973). Self-organization of orientation sensitive cells in the striate cortex. Kybernetik, No.14, pp.85 - 00.

Maskey, S., Dibike, Y.B., Jonoski, A., and Solomatine, D. (2000) Groundwater Model Approximation with Artificial Neural Network for Selecting Optimum Pumping Strategy for Plume Removal, In Schleider and Zijderveld (eds.), *Artificial Intelligence Methods in Civil Engineering Applications*, pp. 67-80.

Mason, J.C., Price, R.K., and Tem'me, A. (1996), A neural network model of rainfall-runoff using radial basis functions. In: *J. of Hydraulic Research, Vol. 34* , pp. 537-548.

Masood-UI-Hassan K., Minns A.W., and Yde L. (1995). Application of artificial neural network to the real time control of hydraulic structures, *In: Advances in Intelligent Data Analysis,* Lasker G. and Liu X. (eds), IIAS Press, pp. 114-118.*atics*,

Mattera D., and Haykin S. (1999) Support vector machines for dynamic reconstruction of a chaotic system. In Scholkopf et al (eds), *Advances in Kernel Methods – Support Vector Learning*, Cambridge, MA, MIT Press, pp. 211-242.

McClelland, J.L., and Rumelhart D.E. (1988) *Explorations in parallel distributed processing; a handbook of models, programs, and exercises*, MIT Press,Cambridge, Mass.

Meer, J.W. v.d., and Franco, L. (1995) *Vertical breakwaters*, Delft hydraulics publications No. 487.

MIKE21 users guide and reference manual (1996) *Hydrodynamic and Boussinesq Wave Modules*.

Minns, A.W., and Hall, M.J. (1996) Artificial neural networks as rainfall-runoff models. *Journal of Hydrological Science*. Vol. 41, No. 3, 399-417.

Minns, A.W. (1998). *Artificial Neural Networks as Subsymbolic Process Descriptors*, PhD thesis, Balkema, Roterdam,

Minns, A.W., and Babovic, V. (1996) Hydrological modelling in hydroinformatics contest, *in Distributed hydrological modelling*, Abbott, M.B., and Refsgaard, J.C. (eds.), Kluwer Academic, Dordrecht, pp. 297-312.

Minoux, M. (1986) *Mathematical Programing: Theory and Algorithms*. Chichester Wiley-Interscience.

Muller, K.R., Smola, A. Ratsch, G., Scholkopf, B., Kohlmorgen, J., and Vapnik, V. (1997) Predicting time series with support vector machines. *In proc. of International Conference on Artificial Neural Networks*, Springer Lecture Notes in Computer Science, pp. 999.

Mynett, A.E. (1999) *Art of modelling*, inaugural address on the occasion of the acceptance of the chair of professor of environmental hydroinformatics. IHE-Delft, The Netherlands (private communications, June 1999).

Norgaard, M., Ravn, O., Poulsen, N.K., and Hansen, L.K. (2000) *Neural Networks for Modelling and Control of Dynamic Systems*, Springer, UK.

Omlin, C.W., and Giles, C.L. (2000) Symbolic Knowledge Representation in Recurrent Neural Networks: Insights from Theoretical Models of Computation, in Cloete and Zurada (eds.), *Knowledge-Based Neurocomputing*, MIT Press.

Osuna, E., Freund, R., and Girosi, F. (1997) An improved training algorithm for support vector machines. *In Proc. of the IEEE Workshop on Neural Networks for Signal Processing VII*, New York, 276-285.

Principe, J.C., Euliano, N.R., and Lefebvre, W.C. (2000) *Neural and Adaptive Systems: Fundamentals through Simulations*, John Wiley, New York.

Saunders, C., Stitson, M.O., Weston, J., Bottou, L., Schölkopf, B., and Smola, A. (1998) Support Vector Machine - Reference Manual, *Royal Holloway Technical Report CSD-TR-98-03*, published by Royal Holloway University press.

Scholkopf, B. (1997) *Support Vector Learning*, R. Oldenbourg, Munich.

Sivapragasam, C., and Liong. S.Y. (2000) Improving Higher Lead Period Runoff Forecasting Accuracy: The SVM Approach. *12th Congress of the Asia and Pacific Regional Division of the IAHR*, Bangkok, Vol. 3, pp. 855 – 861

Sivapragasam, C., Liong, S.Y., and. Pasha, M.F.K. (2001) Rainfall and Runoff Forecasting With SSA-SVM Approach, *Journal of Hydroinformatics,* Vol. 3, No. 3, pp. 141-152.

Smith, J., and Eli R.N. (1995) Neural Network models of rainfall-runoff process. *Journal of water Resources Planning Management,* No. 121, pp. 499-509.

Smola, A. (1996) *Regression Estimation with Support Vector Learning Machines*, Master's Thesis, Technische Universitat Munchen.

Solomatine D.P. (1999) Two strategies of adaptive cluster covering with descent and their comparison to other algorithms, *Journal of Global Optimization,* vol. 14, No. 1, pp. 55-78.

Solomatine, D.P. and Avila Torres, L.A. (1996). Neural network approximation of a hydrodynamic model in optimizing reservoir operation, In: *Proceedings of the second international conference on hydroinformatics*, Zurich. pp. 201-206.

Taylor, J.G. (1996) *Neural Networks and Their Applications*, John Wiley & Sons, NY, USA.

Tefas, A., Kotropoulos C., and Pitas, I. (1999) Enhancing the performance of elastic graph matching for frontal face authentication by using support vector machines, *Proc. of the International Workshop on support vector machine theory and applications (ACCAI '99)*, Chania, Greece

Turing, A.M. (1952) Programming a digital computer to learn, *from the Philosophical Magazine,* (Ser. 7, Vol. XLIII, Dec. 1952)

Vapnik, V. (1998) *Statistical Learning Theory*, Wiley, New York.

Vapnik, V., and Chervonenkis, A. (1974) *Theory of Pattern Recognition [in Russian],* Nauka, Moscow.

Vapnik, V (1979) *Estimation of dependencies Based on Empirical Data* [in Russian] Nauka (English translation Springer Verlag, 1982)

Wang, L.-X. (1992) Fuzzy systems are universal approximators, *In Proc. IEEE Int. Conf. On Fuzzy Systems*, pp. 1163-1170, San Diego, USA.

Whigham, P.A., and Crapper, P.F. (2001) Modelling Rainfall-Runoff using Genetic Programming, *Mathematical and Computer Modelling*, Vol. 33, pp. 707-721.

Williams, R. J. and Zipser, D. (1995) Gradient-based learning algorithms for recurrent networks and their computational complexity. In: Y. Chauvin and D. E. Rumelhart (eds.) *Back-propagation: Theory, Architectures and Applications*, Hillsdale, NJ: Erlbau

Zadeh, L.A. (1973) Outline of new approach to the analysis of complex systems and decision processes, *IEEE Trans. Systems, Man, and Cybernetics*, Vol.1, pp. 28-44.

Printed and bound by CPI Group (UK) Ltd, Croydon, CR0 4YY

23/10/2024

01777685-0012